Mobility Reimagined: My First Decade at the Crossroads of Change

Mobility Reimagined: My First Decade at the Crossroads of Change

Umar Zakir Abdul Hamid

400 Commonwealth Drive
Warrendale, PA 15096-0001 USA
E-mail: CustomerService@sae.org
Phone: 877-606-7323 (inside USA and Canada)
         724-776-4970 (outside USA)
Fax: 724-776-0790

Copyright © 2025 SAE International. All rights reserved.

No part of this publication may be reproduced, stored in a retrieval system, transmitted, in any form or by any means, electronic, mechanical, photocopying, recording, or otherwise, or used for text and data mining, AI training, or similar technologies, without the prior written permission of SAE. For permission and licensing requests, contact SAE Permissions, 400 Commonwealth Drive, Warrendale, PA 15096-0001 USA; e-mail: copyright@sae.org; phone: 724-772-4028

Publisher
Sherry Dickinson Nigam

Product Manager
Amanda Zeidan

Production and Manufacturing Associate
Michelle Silberman

Library of Congress Catalog Number 2025944284
http://dx.doi.org/10.4271/9781468607659

Information contained in this work has been obtained by SAE International from sources believed to be reliable. However, neither SAE International nor its authors guarantee the accuracy or completeness of any information published herein and neither SAE International nor its authors shall be responsible for any errors, omissions, or damages arising out of use of this information. This work is published with the understanding that SAE International and its authors are supplying information but are not attempting to render engineering or other professional services. If such services are required, the assistance of an appropriate professional should be sought.

ISBN-Print   978-1-4686-0764-2
ISBN-PDF   978-1-4686-0765-9
ISBN-epub   978-1-4686-0766-6

To purchase bulk quantities, please contact: SAE Customer Service

E-mail: CustomerService@sae.org
Phone: 877-606-7323 (inside USA and Canada)
          724-776-4970 (outside USA)
Fax: 724-776-0790

Visit the SAE International Bookstore at books.sae.org

Contents

Acknowledgements xi
Voices from the Global Automotive Community xiii
Disclaimer xvii

Part I - Motivation

Chapter 01 - Why This Book (The Summary)
Who Am I 4
    One More Disclaimer 8
Why My Experience Is Unique in a Nutshell 8
The Tone 9
Value 9
    Value for Employers 10
    Value for Employees 11
    Value for General Readers 12
The Objective 13
Summary 14

Chapter 1.5 - My Story: Details
Where Am I Standing Now 17
My Humble Start: The Starting Point 20
Moving Toward Product Management 22
Strategy, Marketing, and Business Leadership 24
Global Experience 33
    Japanese Automotive 33
    EU Automotive 34
    ASEAN Automotive 36
    US Automotive 36
    China Automotive 37

Key Achievements 38
Closing Thoughts 39
    Updates on July 2, 2025 39

Part II - Evolving Industry and Career Landscape

Chapter 02 - The Changing Landscape of the Automotive Industry

Changes that Happened since 2015. 45
The Foundation Shift since 2015 45
Accelerating under Pressure since Post-COVID-19 2020+. 49
External Societal and Generational Factors: Cultural Shifts
and the Rewiring of the Global Order. 52
Summary 55

Chapter 03 - Challenges and Productization in Automotive Mobility Software

The Complexity of Transformation. 57
The All-Around Challenges in Productizing the Automotive Mobility
Software-Driven Landscape. 60
    Software-Defined Products Require a Different Mindset and Way
    of Leadership 61
    Attracting, Retaining, and Managing the Right Talent. 64
    Ensuring Transparency across the Organization to Provide
    a Clear Vision and Strategy. 67
    The Need for Standardized Taxonomy between Different
    Organizations and Functions 69
Conclusions. 71

Chapter 04 - Evolving Career Landscape and Multipotentiality in Automotive Mobility

Understanding Multipotentiality 76
Personal Journey: How Diverse Skills Shaped My Career. 78

The Role of Multipotentiality in Automotive Mobility 79
Requirements for a Future-Proof Team. 80
Criteria to Nurture Your Multipotentiality for Future Relevance 81
Conclusion 82

Part III - Lessons Learned and Reflections

Chapter 05 - A Decade in Automotive Mobility: The Role of Multipotentiality

Key Transitions, Milestones, and Turning Points 88
    From Tech Contributor to Team Lead. 88
    Pivot to Strategic Innovation (ZEEKR). 91
    Global Expansion Leadership (PROTON). 93
Key Moments, Lessons, and Pivots That Shaped My Professional
and Leadership Journey. 96
The Strategic Advantage of Multipotentialites
in Global Mobility 98

Chapter 06 - Transdisciplinary Strategic Leadership: Integrating Technical Expertise, Global Perspectives, and Lifelong Learning

Technical Prowess and Lifelong Learning
for Competitive Advantage 102
Shifting Mindsets: From Projects to Products and Embracing
International Experience. 104
Startups and Corporate Perspectives: Balancing Agility and Stability . . . 106
    The Hidden Power of Startups: Lessons from the Trenches 106
Writing as Your Public Portfolio: Building Reputation and Thought
Leadership 108
    Chase Quality and Impact, Not Mere KPIs. 108
    Take Control of Your Narrative. 109
The Importance of Networking: Lessons from My Journey 110
    Own Your Narrative: Get Involved Beyond the Day Job. 112
Frontier and Emerging vs. Developed Market Strategy—Lessons
You Only Learn by Doing It 113

When Fast Is Not Enough: The Power of Strategic Pauses During
Global Growth and Transformation 115

Building Global Impact: My Lessons on Growing Beyond
the Conventional Path 117

The Power of Networking and Side Projects
for Career Growth 118

Enabling Global Scale Starts with Internal Transformation 119

Coaching the Next Generation 121

My First Decade in Automotive and Mobility Software: Toward
Software-Defined Vehicles—A Brief Remark 123

Lessons for Managers and Recruiters 124

Summary and Conclusion—Staying Relevant, Shaping Your Own
Storyline 125

Part IV - Key Actions and Importance of Multipotentiality

Chapter 07 - Embracing Multipotentiality and Building a Career That Grows With You

Why Now Is the Best Time to Embrace Multipotentiality 130

Building Scalable Career Systems: The Key
to Long-Term Success 132

    Feedback Loops from Mentors, Executives, and Peers 132

    Brand Building Through Public Speaking, Thought Leadership,
    and Visible Project Delivery 133

    LinkedIn: Visibility Beyond Connections 133

    Preparing for the Long Game 134

Challenges Ahead: Why Staying Relevant
Is a Moving Target 134

Summary 135

Chapter 08 - The Future—Beyond the Horizon of Today

Chapter 09 - This Chapter Doesn't Matter, but the Publisher Said I Need One

Chapter 10 - It Is a Fast-Changing World—Only by Collaboration and Diversification of Skills Can We Weather the Storm

Life and Mindset Changes 146
Evolution of the Book. 146
Book Recap. 147
Final Reflections: The Importance of Future and Mental Health 148
Building on 2024—Readiness for 2025 and Beyond 149
    2024 Highlights 149
    2024 Knowledge Strengthened 150
    2025: Non-Negotiables and Aspirations 151
Looking Ahead 151
Final Reflections: A Decade Completed, A New Journey Begins 151
Index 153

Acknowledgements

From changing continents to shifting job roles, I have always remained rooted in the theme of transformation. This book is for those seeking true meaning in their careers and aiming to make a lasting impact in their daily work. Yes, it is my story, but I have learned the most by listening to other people's stories, especially the interesting ones. I believe you will benefit in the same way. Please share this book with your colleagues and friends, promote it, and leave a positive review if you find it valuable.

A huge thank you to Cipawing for pushing me to finish this book. I told you I could do it, but still, thanks. *#ytjt*. My heartfelt gratitude also goes to Sherry Dickinson Nigam and Amanda Zeidan from the SAE publication team, and an honorable mention to the late Barbara Wendling and Linda DeMasi. Your unwavering support in my career journey has meant the world to me.

Thank you to all the mentors and colleagues who have guided me, whether directly or indirectly, over the years. I would love to name you all, but I'm afraid I might unintentionally miss someone, and I wouldn't want that. Please know that you are all part of the reason I am who I am today. Some of you I have never met in person; we connected through email, LinkedIn, or other means. I hope one day to meet you in person.

Finally, I want to dedicate this book to my boy and girl, and to any future child who may one day come into this world and into my life. This book is for you too. I hope that one day you will read it, feel proud, and be inspired. I want you to know what your father has been working on all these years, especially during our time in the Nordics, when you were smaller, and when we started our lives again in Southeast Asia. It has not been easy, but I am trying my best to become the best version of a dad for you. I hope you will see this and understand it. I love you so much, and I wish you every success in your future life.

I also hope this book benefits those who read it. It was written in different parts of the world. The lessons come from many parts of the world. I believe it will be read by readers worldwide as well. But one point remains: regardless of what we chase professionally, please remember to be kind to each other. We are all human. People will remember how we treat them. I remind myself of this as well.

If you would like further expansion of this book, whether through workshops, training, or other services, please contact me and we can arrange something further.

Yours truly,
Umar Zakir Abdul Hamid
June 8, 2025
Somewhere in Southeast Asia, 1508H

Voices from the Global Automotive Community

This book is written not only to share my story but also to highlight a key perspective: the ongoing transformation in the automotive, mobility, and related industries is reflected in the rise of diverse demographics within the global automotive talent pool. Yet, the voices of these diverse groups and minorities are often unheard on the global stage.

Having worked extensively in the automotive and mobility sectors, I have built a broad personal network spanning both the Global South and Global North. This network covers not only developed markets but also frontier and emerging markets.

To co-create impact with this book, I have crowdsourced forewords from professionals of diverse backgrounds worldwide. These contributors explain why this topic matters to them. I carefully selected the voices featured here, each bringing unique backgrounds and perspectives. I hope their insights add weight to this book, especially from the perspectives of diversity and real-world experience.

Perspective I: Omar Desouky

Product Planning Supervisor, Proton Egypt

Ezz Elarab Automotive Group

Egypt

The automotive industry is evolving faster than ever, with innovation, localization, and strategic foresight becoming essential to long-term success. Working on the ground in a fast-paced market like Egypt, I've had the chance to see just how important those elements are and to experience firsthand the impact of strong leadership and clear vision.

I've had the pleasure of working closely with Dr. Umar Zakir Abdul Hamid on shaping Proton's product direction in Egypt. Our collaboration was especially meaningful during the launch of the Proton Saga; the first locally assembled Proton model in the country. It wasn't just about introducing a new car; it was about making the brand more accessible to Egyptian consumers and laying the groundwork for Proton's growth across the region.

Umar's leadership at Proton HQ was instrumental in this. He brings a unique blend of strategic insight and practical understanding that helped bridge the gap between global objectives and local market needs. The success of the Saga was just the beginning, there are more innovations in the pipeline, all designed to elevate the brand and strengthen its presence not just in Egypt, but throughout the Middle East and North Africa.

Having worked with Umar and followed his insights over the years, I see this book as a natural extension of the way he thinks forward-looking, grounded in real world experience, and always connected to a bigger purpose. He doesn't just talk about the future of mobility; he works to shape it in meaningful ways.

At a time when our industry is navigating rapid shifts in technology, consumer expectations, and sustainability, this book couldn't be more timely. I'm confident it will offer readers valuable perspective, practical insight, and a fresh way of thinking about the road ahead.

Perspective II: Jessica Swafford, MBA

University of Oxford, Jesus College

USA

Umar and I met while I was studying at Oxford, and he was a guest panelist at an automotive event. We both worked in the auto industry, and he later wrote a submission for my *Oxford Strategy Review* column called, *"The Coming Future: Mobility Disruption."* Years later, we co-wrote a submission for *Automotive News* titled *"Customer*

Centricity: A Necessity for Automotive and Mobility Software Product Development."

These pieces discussed the need to be nimble in the workforce and above all else, to champion the customer. As Umar has been preaching these important themes for years, it is no surprise that he has penned a book that includes these topics. Umar carries a wealth of knowledge across the automotive spectrum including vehicles and technology. Few people have such varied experience in an industry rife with nuance. I am particularly excited to read his thoughts from an international lens. Having lived on multiple continents, he is no doubt an expert in the ever-changing, global transportation industry.

Perspective III: Khairul Akmal Zulkepli

System Architecture Manager (Automotive Software, Autonomous Driving)

eMoovit, Malaysia

Congratulations to you, the reader, for picking up this book. I believe there's something in it for everyone—if not today, then perhaps in the future.

I first crossed paths with Dr. Umar in 2015, when he was pursuing his PhD and I was working as an engineer. The journey of a PhD candidate is no small feat; it's a demanding path filled with complex topics and limited resources. Yet what stood out to me then, and remains true today, is Dr. Umar's unwavering commitment to his work and his deep belief in its global impact—particularly in advancing the autonomous vehicle and robotics industries. And indeed, his work has made a difference.

In a world where anyone can publish, trust has become a precious commodity. Knowing Dr. Umar personally, I can say that trust in his work came naturally to me. Over the years, I've followed his contributions—both formally and informally—and they've consistently provided valuable insights into the challenges of next-generation mobility. His perspectives have helped me navigate the autonomous vehicle space and unlock opportunities in this growing, dynamic industry.

Lastly, I believe Dr. Umar's reflections on the importance of diversity are both timely and necessary. In today's interconnected world, where geographical barriers are fading, we must harness the value of diverse voices and perspectives. His call to view diversity not as a challenge, but as a strength, is a message that resonates strongly—and one we should all embrace to build a more inclusive, innovative future.

Perspective IV: Ameer Faiz Mahfuz

APAC & Oceania Regional Sales Lead, PROTON International

Malaysia

This book is very timely because the automotive industry is changing fast, especially with the move towards electric vehicles (EVs) and future mobility. It gives a clear and real view of what we need to grow in this industry—not just technical skills, but also leadership, global thinking, and how to follow new trends.

I first met Dr. Umar when he joined Proton. Before that, he was with ZEEKR, one of the well-known global EV companies. I knew from the start that he would bring his valuable experience from ZEEKR into Proton, and he did. From our discussions, I always found him very open to share ideas. He often talked about how European and global carmakers solve their problems and how we can apply the same thinking to improve our own products and systems.

For young people in the automotive industry like me, this book is very helpful. It shows that the future will be led by those who are willing to learn, adapt, and bring fresh ideas. It is also a good guide for those already in the industry who want to keep up with the EV transformation.

Dr. Umar shares real-life stories and global experience that teach us about innovation, diversity, and how to move forward. This book will motivate and guide anyone who wants to succeed in the next chapter of the automotive world.

Disclaimer

This book is a personal reflection of my career journey and professional experiences. The views, opinions, and insights expressed herein are solely my own and do not represent or reflect those of my current or former employers, their affiliates, partners, or any related entities. No portion of this work should be interpreted as an official statement or position of any organization with which I am or have been associated. I am writing this book in my individual capacity as an industry professional and leader, with the intent of sharing personal learnings and perspectives.

Part I

Motivation

Chapter 01

Why This Book (The Summary)

I signed the contract for this book more than a year ago and have written more than ten different initial versions. But for some reason, none of them feel right or sit well with me. It is not what I want to share with the world.

I do not want to write a book just for the sake of writing or publicity. And every time I delay and rewrite, I feel like a completely different person from the last "stop." Working in one of the fastest-transforming industries—automotive—my experiences and capabilities evolve every month (I'm not bluffing).

But I need to finish this book. I want to share my story. If not now, it will not happen. And even if it does happen in the future, I will be telling a different kind of story to a different kind of audience. Right now, my focus is on the audience of this book.

For this one, I want it to be special—something rooted in knowledge-sharing, community-building, and democratizing access to success. Inspiring minorities, aspiring youngsters, and paving the way for underserved talent demographics.

Beyond that, it marks a milestone, signifying more than a decade of my career in mobility.

Some may ask, "Why are you so full of yourself? Why write a book about your own story? What's so special about it?"

The quick answer: As a physically brown guy who has worked in multifaceted areas of the automotive industry across the EU, China, Asia, and Japan—holding strategic positions in different organizations and winning a few awards in next-gen mobility—before even reaching 35, I have stories to tell that will benefit younger talents out there.

Who Am I

I am currently heading International Market Strategy, Expansion Strategy, and Global Product at a Southeast Asia-headquartered subsidiary of a major global automotive conglomerate based in China. Since 2021, I have been with the conglomerate across two of its brands and have worked behind the scenes, collaborating with several other sister brands.

On paper, this may seem quite different from my previous work in the tech scene. While it is indeed a different journey, it remains closely related to what I was doing before—bringing emerging technologies to the world. Whether it's connected vehicles, autonomous driving (AD), electric vehicles (EVs), shared mobility, entry-level internal combustion engine (ICE) models integrating connected infotainment head units (IHUs), or even leading the global expansion of a traditionally domestic legacy company, the core mission remains the same: transformation.

To productize emerging technologies, we need diverse perspectives. And to gain diverse perspectives, we must first acknowledge a fundamental issue: a lack of diversity. Increasing diversity requires listening to different stories, and I believe mine is worth sharing.

Now, let's rewind—not too far, but far enough to provide the bigger picture of why I believe my story matters. I have summarized below my career journey so far (sorted from latest to oldest). For clarity, I have paraphrased some titles to better reflect what I actually did rather than just the official title, because "Researcher" (for example) does not fully capture the scope of my work.

- Head of International Strategy, Market Expansion, International Business Governance, and Global Product) (Present) | PROTON (Part of GEELY Group), Malaysia
- Senior Lead Strategist (reporting to VP and SVP) (2021–2023) | China Euro Vehicle Technology AB (ZEEKR EU HQ, GEELY Group), Sweden
- Product Owner, Engineering Team Manager (Autonomous Vehicle Planning & Control Product Software Development) (2018–2021) | Sensible 4, Finland
- Autonomous Vehicle Scientist (2017–2018) | MooVita, Singapore & Malaysia
- Visiting Researcher (Vehicle Active Safety) (2016) | Smart Mobility Research Center, Tokyo
- PhD Researcher (Autonomous Vehicle Control, Advanced Driver Assistance Systems) (2014–2017) | Universiti Teknologi Malaysia
- Product Development Executive (2014) | IRIS Corporation, Malaysia

Quick disclaimer: the book is about my career journey and is not affiliated with any of these organizations.

Now, let's continue.

I have worked in various sectors of the automotive industry: autonomous vehicles (AVs), robotics, EVs, connected vehicles, legacy ICE vehicles, and shared mobility. Beyond that, my experience spans continents and different positions. From the bottom to the top, I have been exposed to many elements of the industry. I won't share too much in this chapter, as it is meant to provide an overview of the book. More details will follow in the next chapter.

But the key here is that my unique experience consists of the following:

1. I have worked across the end-to-end functions of the next-gen mobility business, covering pure R&D (academia, startup, and industry), product management, software product ownership, business strategy, and strategic marketing—interfacing directly with the customer.

2. My global perspective and multicultural dexterity define my experience. I have worked in Asia (Japan), ASEAN (Singapore, Malaysia), and the EU (Finland, Sweden), with close connections to the US automotive community through active participation in SAE International committees. Additionally, I have worked with a Chinese-based automotive conglomerate. I have worked with different automotive networks—Japanese, ASEAN, Chinese, European, and North American—each with unique approaches to connected, autonomous, shared, electric (CASE) mobility. Furthermore, my exposure to international business—spanning the Global North (EU, US) and now the Global South (developing, frontier, and emerging markets with PROTON)—has given me a full-spectrum perspective of the end-to-end automotive business. Living in the Nordic region for more than half a decade also exposed me to a distinct kind of leadership experience, and I observed how such regions with low populations can produce a lot of innovation outputs. This exposure has given me a deep understanding of varying market dynamics, consumer behaviors, business notions, and cultural sensitivities, shaping me into a well-rounded automotive business professional.

3. I have been at the forefront of digital transformation in mobility and industrialization. AVs are not new, but I was fortunate to start in 2014, just as the hype of investments for AV commercialization was ramping up. I also witnessed the realization phase (the decline of AV hype) during my transition to EVs, coinciding with mass electrification efforts like COP 26 that took place in Glasgow, which sees commitments from carmakers related to zero-emission vehicles and the rise of BYD.[1,2]

4. I have navigated and worked through the full lifecycle of next-gen mobility product and market strategies—from conceptualization in R&D (GEELY R&D Centre, formerly CEVT, before becoming ZEEKR

---

[1] https://impact.economist.com/sustainability/net-zero-and-energy/cop26-november-10th-and-11th-its-time-to-be-bold

[2] Sousa, C.D.M.D., 2024. Electric Vehicle Manufacturer BYD Expansion into Europe: A Case Study (Master's thesis).

Tech Europe), to market entry (ZEEKR EU with ZEEKR 001 and ZEEKR X from China), and global business expansion strategies (PROTON, as part of the GEELY Group Family-related companies).

5. I have a proven portfolio of commercial acumen. Beyond my work in CASE mobility, I have delivered tangible results. When I joined Sensible 4, the company had fewer than ten people; by the time I left in 2021, it had grown to over 100. As a key figure in this rapidly growing auto-tech startup, I contributed to securing over €20M (funding and revenues combined). Additionally, I have strategized and contributed to the commercialization of digital service products at CEVT (now ZEEKR Technology Europe). I also drove market entry for ZEEKR EU and international expansion for PROTON in its ASEAN HQ. I currently lead the foundational and strategic efforts for PROTON's NEV Export program—our organization's first such initiative and one of only two native ASEAN companies to do so (the other being VINFAST in Vietnam). While writing this book, we had just launched our first EV in Nepal—another milestone in our history—where I led the foundational work from scratch.

6. As of November 2024, I have been frequently invited to top conferences across academia, startups, and industry. Most recently, in October 2024, I was invited as a keynote speaker (online) at one of the prestigious IROS workshops, organized by Prof. Zolotas of Cranfield University (UK), held in the UAE. This demonstrates that my expertise is recognized and valued across different organizations.

7. My involvement in market entry and product management, combined with strong technical knowledge, has also exposed me to regulatory and governance expertise. I served as the subject matter expert for GDPR compliance at CEVT and ZEEKR EU (an additional role that I had in the organization during that time due to the rapid growth of ZEEKR expansion ambition requiring additional resources), and have independently served as Secretary of SAE's Cooperative Driving Automation Committee since late 2020. This reflects my deep understanding of regulatory landscapes, ensuring that the strategies I develop, lead, or contribute to are not only innovative but also globally compliant – enhancing scalability and credibility.

My growth has largely been driven by two factors: (i) timing and (ii) taking ownership when no one else did, especially during periods of rapid change. An example of this was the recent tariff situation between the West and East economic bloc due to the second Trump presidency in 2025.[3] This situation forced me to step in and lead our organization in understanding its impact on our international export strategy.

In a sense, these events provided me with opportunities for growth. These are the reasons why I believe my experience is worth sharing. For brevity, the details of my stories will be provided separately in their respective chapters.

And it is important to highlight that the reason why I want to write this book is to document all these, so that I myself will not forget. Like one of the professors at my PhD alma mater was saying, "What better way to prove that you have done something apart from writing?"

One More Disclaimer

I believe in diversity, equity, and inclusion (DEI), but merit should also be a key factor. Having worked in different countries, I've experienced being sidelined at times due to the irony of DEI efforts. While racism is one issue, other biases, such as ageism and sexism, also exist. As a disclaimer, this book aims to offer new perspectives and may touch on topics some consider controversial. My intention is not to create controversy but to highlight these issues based on my personal experiences.

Why My Experience Is Unique in a Nutshell

I was in China a few days ago visiting our co-parent (November 2024) and witnessed disruptions that remain theoretical in other countries but are already a reality in China, especially in the automotive sector. The East's economic powerhouse is leading the transformation. As the industry undergoes rapid change, fresh perspectives are essential. Traditionally, our field has been led by men in their 50s to 70s, white- and gray-haired, from the West or East Asia (excluding China).

---

3   https://en.wikipedia.org/wiki/Trump%27s_Liberation_Day_tariffs

However, with software now in the picture, hoodies are replacing neckties. So, is someone with a brown physical appearance, a professional presence in the EU, JP, and ASEAN, working with a CN-based conglomerate, experienced in startups and scale-ups, leading projects with EU exposure, a strong academic reputation, a guest speaker at automotive and future mobility events, an author, SAE secretary for CDA since 2020, a recipient of multiple awards, and holding an executive MBA (EMBA) and PhD not unique?

The Tone

Anyway, you might have noticed—the tone is quite simple.

I want this book to feel personal but not like fiction, professional but not too uptight. I want it to read like a sharing session. To reach as many people as possible, I'll use simple words—the best I can. The goal is to present a storyline that says, "Hey, change is happening. This is my perspective. Maybe you can take something from it and apply it to your future planning—whether in your career, profession, organization, or beyond."

I also want this book to be as economical as possible, so I've made the decision to minimize the number of photos, or possibly exclude them entirely. My primary goal is for people from all countries to be able to afford it. Initially, I considered including a lot of photos and aesthetic embellishments, but after careful consideration, I realized that affordability is paramount.

Rather than writing 19 chapters, which could drive the price up, I've opted for simplicity. My hope is that this will make the book accessible to the masses, ensuring that the valuable insights are within reach of anyone who wants to read this book.

Value

In this book, I expect to provide strategic value to three main reader categories: employers, employees, and general readers. I believe that, due to the knowledge gap, many topics that are discussed in this book are not fully recognized in hiring, attracting, and managing talent.

Value for Employers

Around the pandemic, the automotive industry saw major hiring, with many people being brought in and massive investments made.[4,5] However, by 2022, following a reality check post-pandemic, layoffs ensued as redundancies emerged.[6]

Was it truly redundancy or a failure to understand future requirements? A failure to manage this new demographic of talent? A failure to grasp the direction we want to go? Or all of the above—trying to steer into a third-gen strategy with a second-gen organization, built by first-gen leadership?

To me, this is a classic example: it's not about how much you hire, but who you hire. Is 20 years of experience really the right criterion for an EV head program when the major shifts in the EV and AV industries have only taken place approximately over the past seven years?

So, who should we actually hire?

In new areas with unclear terminology and taxonomy, such as digital marketing, what does "digital" mean in the automotive industry? Does it refer to digital communication or is it related to software? Should the person have a technical background, pure sales, or business? Or should it be a full bridge of both? How do we find these people? Are all companies using the same taxonomy?

This is what I am trying to bring to the table. By sharing my stories, which are not unique to me alone, I hope to help employers understand the voice of the "unheard."

---

4   Dakić, P., Stupavský, I. and Todorović, V., 2024. The Effects of Global Market Changes on Automotive Manufacturing and Embedded Software. *Sustainability*, 16(12), p. 4926.
5   Hamid, U.Z.A., 2022. *Autonomous, Connected, Electric and Shared Vehicles: Disrupting the Automotive and Mobility Sectors*. SAE International.
6   El-Deeb, A., 2023. The First Tech Layoff Wave After Years of Hypergrowth: How This Affects the Industry?. *ACM SIGSOFT Software Engineering Notes*, 48(1), pp. 4–5.

Value for Employees

One core aspect of this revolution is the disruption and democratization of opportunity. Previously, this sector was dominated by certain demographics in specific countries. Now, CASE mobility integrates new technologies from various fields into the automotive industry's main product line.

Software-defined battery EVs (BEVs) place software at the core of the product, requiring talent and business strategies from software-driven industries. This shift creates opportunities for new demographics to enter the automotive industry.

Being no. 1 in car sales does not equate to being no. 1 in software development, which is why many legacy carmakers struggle with this transition, while new players adapt quickly.

Regarding talent diversification, hiring, and attraction, I have written extensively on this topic across different outlets. Examples are below:

1. *Autonomous, Connected, Electric and Shared Vehicles: Disrupting the Automotive and Mobility Sectors*, https://www.sae.org/publications/books/content/r-517/
2. Talent: Overlooked Key to ACES Mobility Productization, https://www.eetimes.eu/talent-overlooked-key-to-aces-mobility-productization/
3. Challenges of Next-Generation Automotive and Mobility Innovations' Productization and Industrialization, https://www.sae.org/blog/gp-umar-zakir-abdul-hamid-next-generation-mobility-productization-industrialization
4. Challenges of Complex Software Development for Emerging Technologies in the Automotive Industry: Bridging the Gap of Knowledge Between the Industry Practitioners – SAE WCX 2022, https://www.sae.org/publications/technical-papers/content/2022-01-0109/

For this book, I am going full force, deep-diving into this topic to prepare not only employers but also employees. The goal is to equip ourselves with the right skills needed to survive and remain relevant in the future job sector. To some extent, I will not explicitly state what should be done, but based on my experience, you will grasp the gist and notion of how to stay relevant. After all, it is better to share examples than to focus solely on theories.

Value for General Readers

When I was in Finland during the coronavirus (COVID-19) pandemic, I spent a lot of time reflecting on the automotive industry. While actively deploying public projects and prototypes for government agencies in the EU and private customers with our AV technology, I often asked myself: Is this really the future? I mean, it is the future—but is it the near, feasible future? My concerns were grounded in hands-on experience, not just fanboy arguments defending or criticizing future mobility.

During this time, I reached out to my network, including Mari Suoheimo, an expert in service design from Finland, who is now an Associate Professor at The Oslo School of Architecture and Design (AHO), Norway. Through LinkedIn, we connected and eventually co-edited a book, featuring contributions from top service design experts worldwide, including those from MIT and other leading laboratories in the service design field.

As I was finishing this book, Jaguar (or should I say JaGUar?) launched a campaign to rebrand and reframe itself to stay relevant.[7] In parallel, they halted car production for two years.[8] Their main agenda? A complete reboot. To me, this made perfect sense. To remain relevant, they needed to focus on a viable business audience rather than risking the fate of past industry giants that failed to adapt to disruption. Will they succeed? No one knows. But will they fail if they do nothing? History suggests so.

My point is no one truly knows what the future holds. What we do know is that the industry and the world are transforming. The rapid adoption of software-defined and data-driven technologies requires us

---

7   https://youtu.be/rLtFIrqhfng?feature=shared
8   https://www.bbc.com/news/articles/c7v3m826njno

to prepare for this transition. One way I contribute is by sharing my story, because I am living this transition. I hope this helps aspiring young people in developing countries prepare for the future and inspires them further.

I am also writing this book to advocate for genuine DEI with meritocracy on a global scale. Reaching where I am today wasn't easy. I moved across countries and continents to gain perspectives that weren't available where I came from. But I believe integrating new perspectives is key to driving business innovation.

The Objective

This might be repetitive, but to be clear—based on my diverse career experience across different continents and countries—this book is built on four main pillars:

1. Sharing my story to inspire and motivate young generations from diverse backgrounds, proving that career growth like mine is achievable.
2. Providing an overview of industry disruptions, particularly in the automotive sector, from the perspective of career evolution.
3. Offering insights on the future career landscape to organizations—academic, industrial, and public—helping agencies create better job opportunities and enabling companies to design their organizations to align with future business directions.
4. Serving as my portfolio, as only I fully understand the depth of my experiences across different countries.

I hope this book successfully achieves its goals.

From a small automotive market in Southeast Asia to influencing billion-dollar potential strategic planning in Europe and Southeast Asia for an automotive conglomerate based in China, my journey is not typical, but it's instructive and can be replicated by aspiring young professionals.

I have transitioned from a deep engineering role toward a more customer-facing, strategic business leadership role that defines my current organization's global expansion. And that's why I believe that this book can help others navigate the nonlinear and, most of the time, uncertain career growth in a fast-evolving field like automotive mobility.

But for me, this knowledge can also be applied to other industries. I believe this book can guide future professionals navigating the complexity of CASE mobility, digitalization, and cross-regional leadership.

And finally, I believe it can also serve as a case study or example to promote more DEI in the workplace—to show that we cannot put a ceiling on someone's career growth just because of their skin color.

To the readers, I hope this will inspire you to reflect on your career growth journey.

Summary

This book is written amid a complicated personal crisis that forced me to move back to my home country. Fortunately, I am still working with the same parent group, GEELY, and I consider myself lucky that ASEAN and emerging markets are at the forefront of disruptions in CASE mobility, especially in electrification. These disruptions offer perspectives that developed markets may not see. Looking back, this transcontinental move has broadened my perspective on the automotive industry. I am no longer focused solely on developed markets, but also on legacy automotive companies and emerging markets—truly a blessing in disguise.

To advance technology with cross-functional inputs, we need more dialogue. Of course, later chapters will explore this further. But I believe this book will offer fresh perspectives on the technological side, written with transparency, vulnerability, agility, and humanity.

In summary, this book provides artifacts that support the ideas discussed in this chapter, aiming to inspire readers to progress in their careers. I am writing this because no one truly understands the experiences I went through or the knowledge I gained along the way—most of which revolves around soft skills. This book documents that journey.

The next chapter shares my detailed story, including some personal aspects never before made public. But rest assured, automotive readers will still find value. After that, we will dive into the disruptions and the future of mobility careers, one by one.

Happy reading—enjoy, and hopefully, this will motivate you to navigate change with confidence. And of course, no matter how much I strive for perfection, this book is far from it. A career is a learning curve or, as some might say, a real-life business school.

Lastly, to myself: as much as you want to align this book with the daily growth you are having, you cannot. Because (i) this is not a holy book and (ii) that is the beauty of imperfection—life is all about learning and becoming a better version of your current self.

Chapter 1.5

My Story: Details

Where should I start?

Should I begin ten years later—the fancy, LinkedIn-worthy version? Or the less popular one? Well, it doesn't matter. Since I have this opportunity, I'll write a detailed summary of my journey. I believe many unspoken artifacts can be deciphered from it.

After this chapter, we'll dive into its insights in detail. For now, this chapter will focus solely on my career journey so far.

Where Am I Standing Now

As of Q1 2025, this is my LinkedIn profile summary:

Umar Zakir Abdul Hamid, PhD & EMBA, is a globally recognized leader in Next-Gen Automotive Business, specializing in International Business, Market Expansion, Corporate & Product Strategy, and Strategic Market Management, with expertise in Software-Defined Vehicles, Digitalization, X-as-a-Service & CASE Mobility.

As a Business & Corporate Strategist:

- *Co-led €20M+ funding & revenue at Sensible 4, drove EU market entry & innovation at ZEEKR EU (GEELY Group) in digital, connected, and software products, and now leads global expansion strategy and the electrification export task force at PROTON International (GEELY Group).*
- *Developing market entry strategies and product alignment across 18+ countries, integrating Nordic business acumen with ASEAN growth dynamics.*
- *Defines global business strategy, international market entry models (CBU, CKD, SKD), & regulatory pathways for scalable expansion (PROTON).*

As a Market Expansion & Product Strategy Leader:

- *Drives PROTON's global expansion, volume forecasting & scalability, transforming it from a domestic-focused brand into a global player.*
- *Manages product-market fit, pricing, and strategic positioning for NEV & ICE models, ensuring regulatory compliance and profitability.*
- *Developed PROTON's first global NEV export roadmap (BEV, PHEV, HEV, REEV) in collaboration with Geely HQ, ensuring regulatory alignment and competitive positioning.*
- *Drives cross-functional collaboration with global distributors, Geely & internal leadership for seamless market alignment.*
- *$2B+ potential revenue impact, 300× volume growth potential, subject to operational success (PROTON).*

As an Engineering Manager, Tech Lead & Expert:

- *Ten+ years driving advancements in Software-Defined Vehicles, Electrification, Digitalization, and CASE Mobility, shaping global automotive standards.*
- *Led 12 direct reports from 10-nationality experts, overseeing 20 dotted-line reports at Sensible 4 (Finland).*
- *Pioneered ZEEKR Connected & CEVT innovation commercialization strategies in the GEELY EU Ecosystem, advancing autonomous and digital ecosystems with Sensible 4.*

- *Shaped PROTON's global electrification, expansion, and digitalization strategy, aligning with future mobility trends.*
- *Recognized with the SAE Young Industry Leadership Award (2023) and Finnish Engineering Award (2020).*

To Recruiters & Collaborators:

- *Bringing senior leadership expertise in global strategy, product innovation, market expansion, and corporate growth for high-impact roles.*
- *Driving success in a high-paced, rapidly evolving industry, leading transformation amid constant market shifts.*

However, this summary does not fully reflect my journey or justify the necessity of this book—especially for those aspiring to follow a similar path, such as transitioning from academia to industry, moving from industry expert to product, or evolving from product to strategic leadership. A more detailed account is needed. This chapter will provide the full picture, while the next chapters will offer an in-depth deep dive into the context.

And as I write this, I am leading PROTON International's Market Strategy, Expansion Planning, and Global Product across various ICE and new energy vehicle (NEV) models in more than 20 countries. We are in the midst of revamping the company's internationalization, a journey filled with excitement, surprises, and stress. Update: As I do the final proofreading of this book on July 20, 2025, we have successfully registered the incorporation of PROTON International Corporation Sendirian Berhad (Limited), which will act as a stand alone corporate entity to operate the export strategy that I co-architected with our head of division earlier this year. Time flies so fast. I was also the main architect and leader of the business operating model framework that this new company will use in the years to come, co-designed together with other colleagues.

How did I get here again? Let's go. Some of it is my effort. Most of it comes from being part of great teams and having incredible mentors. And the rest? Pure luck and timing—being in the right place at the right time. Yes, as corny as it may sound.

And before you ask, I deliberately labeled this chapter as Chapter 1.5 instead of Chapter 2 to emphasize that while it is not part of the book's main narrative, it is essential for understanding the insights.

My Humble Start: The Starting Point

After spending about six years in Moscow completing my undergraduate studies (Russian Engineering Diplom), I worked for a while before being accepted into the PhD program *at Universiti Teknologi Malaysia's Malaysia-Japan International Institute of Technology (MJIIT), Kuala Lumpur campus.* It was part of an industrial research program involving collaboration between PROTON (pre-GEELY), the Smart Mobility Research Centre in Tokyo, and Universiti Teknologi Malaysia. *The work involved cross-border collaboration between Malaysia and Japan.*

I was part of the pioneering team of researchers building Malaysia's first AV algorithm and software from the ground up—a trailblazing effort. The atmosphere was electric and hyped, pure R&D, with everyone highly motivated. In the middle of the research, I had a stint in Tokyo working with Professor Pongsathorn Raksincharoensak, a close collaborator and, to some extent, a protégé of Professor Masao Nagai, one of the world's top vehicle dynamics professors with strong ties to Japan Automobile Research Institute (JARI).

During this time, I gained deep insights into Japanese R&D methodology and mindset. Beyond work, I also observed Japanese culture and began to understand why their innovation is unique—it is embedded in their lifestyle and perspective on life.

Throughout my PhD, I wrote extensively, submitting ten papers by the time of my viva voce presentation. Working in Japan exposed me to the importance of networking—something Pongsathorn Raksincharoensak subtly instilled in me. I reached out to numerous R&D experts, consulting them for feedback on my thesis ideas, some of which were reflected in my final work.

In the acknowledgment section of my thesis, I recognized these individuals: Hairi Zamzuri (Universiti Teknologi Malaysia), Mohd Azizi

Abdul Rahman (Universiti Teknologi Malaysia), Pongsathorn Raksincharoensak (Smart Mobility Research Center, Tokyo), Yuichi Saito (Smart Mobility Research Center, Tokyo), Mohammad Ali (Volvo Car Group, Sweden), Ugo Rosolia (University of California, Berkeley, US), Ondrej Mikulas (Honeywell Automotive Software, Czech Republic), Ondrej Santin (Honeywell Automotive Software, Czech Republic), Stephanie Lefevre (Mercedes-Benz R&D, US), Yanjun Huang (University of Waterloo, Canada), Tulga Ersal (University of Michigan, US), Muhammad Aizzat Zakaria (Universiti Malaysia Pahang, Malaysia), Mohd Hatta Mohammed Ariff (Universiti Teknologi Malaysia), Volkan Sezer (Istanbul Technical University, Turkey), Xiuyan Guo (Autonomos Lab, Berlin), Konstantin Pushkin (Moscow Aviation Institute, Russia), Toru Saito (Honda R&D, Japan), Shuhaibul Fadly Mansor (Proton Holdings Berhad), and the TAW Team of Vehicle System Engineering iKohza (Malaysia).

This was not just a formality: it reflected my approach to the PhD. I pursued a deep understanding of the subject, unafraid to explore its philosophical aspects. I engaged with experts across different regions, companies, and institutions, treating my research as an opportunity to expand my knowledge beyond academia.

Before I even finished my PhD, an opportunity came in. A team from Singapore, originally from Agency for Science, Technology and Research (A*STAR), wanted to join forces. So, we became the full MooVita—possibly the pioneering cross-country AV team in ASEAN and the first to build an AV in Southeast Asia in-house with commercialization goals.

During this period, I started using LinkedIn actively, connecting with international professionals, and closely observing industry trends. I realized the importance of "going further," especially with encouragement from my mentors.

I remember working late, around 1 or 2 AM in Singapore on a project during that time, questioning myself, whether this was really the way to develop such a complex innovation—or whether there were other ways to bring it to the public. In a sense, I have always had a systemic-level thinking mindset: the ability to see the big picture or, as the business world puts it, thinking strategically.

Around that time, with the birth of my child and other personal developments, I started looking for opportunities in a country where I could achieve better work–life balance. It may sound cliché, but it was a real priority. I even discussed this perspective in my TEDx Talk in Helsinki during the pandemic.[1] Working in the Nordic environment accelerated my growth (more on this later).

Several opportunities and interviews came my way from the EU, Asia, and the US. Then one night, I saw a post about "a team developing autonomous vehicles for extreme weather conditions." Having spent four years in this field during that time, I knew the topic was massive, with immense potential for monetization, research, and growth. So, I sent an application—or rather, an email—to the founders of Sensible 4, Harri Santamala and Jari Saarinen.

Shortly after, Jari responded: "Hey, do you wanna talk?" Or something along those lines. Fast forward three months (or less), my family and I moved to Finland.

To be honest, the salary wasn't high. We lived in a small apartment, but I was drawn to the company's technology and its potential. At that point, my interest was still primarily in the technology. But of course, the move was also about having more time to be a father while focusing on my personal growth, which, in hindsight, was the right decision.

When I joined Sensible 4, the team had fewer than ten people (more or less). It was a true deep-tech garage startup—mechanical engineers, a handful of software experts, and the management team. I was there from the beginning, helping scale the company. Part of the superhero team.

Moving Toward Product Management

After ~1.5 years of building the tech while simultaneously working on EU public projects and proof of concept projects for corporate companies (mostly confidential), we finally secured our first

---

[1] https://www.ted.com/talks/umar_zakir_abdul_hamid_lifelong_learning_in_finland_an_expat_s_view

investment—around €6M.[2] Before that, we also won $1M in the first-ever Self-Driving Challenge Award in Dubai for the startup category.[3]

Just imagine: a small team winning an automotive tech award in its inaugural year, which has now become a prestigious industry event. Every year, we generated encouraging revenue (for a small company of that size), and by the time I left, my role as a key cornerstone member of Sensible 4 had contributed to a collective revenue and investment total of €20M since I joined.

After that first investment, the real turning point came—the moment that shaped me into a global strategic business leader in the automotive world with a technical foundation and cross-functional execution ability.

Little did I know that when the investment arrived, it wasn't just about scaling the technology—it was actually about building the company itself, from scratch to a fully structured organization. No processes? We created them. No HR? We hired one. We were corporatizing the garage-world culture, navigating internal earthquakes along the way.

But here's the thing: those almost 3.5 years of (beautiful) chaos at Sensible 4, in an extremely fast-growing startup, gave me an unparalleled pace of growth that only those who've been through it can truly understand. For outsiders in stable, large corporations, 3.5 years is just business as usual—a mellow timeline with minimal disruption and little personal acceleration.

After the investment, we scaled rapidly—from 20 to 100 people in less than a year, all during the pandemic. In my team alone, I interviewed more than 30 global experts and built a direct-report team of 12 planning and control algorithm software specialists from 10 different nationalities. Including dotted-line reports, I led about 30 engineers across full-stack AV and robotics technology, covering positioning and mapping, obstacle detection and tracking, remote control and connectivity, and low-level embedded systems. I led them as the solution architect and product owner for the planning and control stack for AVs.

---

2 https://sensible4.fi/2020/02/28/finnish-sensible-4-raises-7-million-to-support-expansion-of-autonomous-driving-system-specialised-for-harsh-weather-conditions/

3 https://sensible4.fi/company/newsroom/winner-of-dubai-world-challenge-for-self-driving-transport/

Speaking of diversity, it was not intentional to hire from multiple nationalities, but as I structured the organization, I quickly realized how difficult it was to find all the right talents in one place. That is when it became clear: in emerging industries and technologies, talent is scattered globally—another unspoken challenge of productizing cutting-edge technology.

At the same time, we were still building the company and bringing people in. While doing all this, we began implementing structured software processes: introducing Scrum, Agile philosophy, ISO 26262, SOTIF, and discussions on refactoring code, requirements engineering, and managing technical debt—a lot at once.

What a beautiful earthquake to be in!

We learned the language of the automotive industry, attended conferences, and adapted fast. As more companies entered the space as our customers, expectations shifted. It was no longer just about proving concepts; the market demanded at least pre-production-ready quality.

Good experience. Opportunities don't come often. When they do, take them.

Strategy, Marketing, and Business Leadership

Sensible 4 was one of the fastest-growing startups in the Nordics—possibly one of the most hyped autonomous vehicle companies in the EU at the time. When the investment happened, we didn't just need to scale the company; we were also hit with an unexpected challenge: the pandemic. Some called it COVID-19, while in the Nordics, many referred to it as corona.

During that era, not only were we scaling after securing investment, but we also had to set up processes, documentation, and strategy—all while working to commercialize our AD stack. A blessing in disguise was that I had time to take several online courses in business, including a product management professional certificate from the University of Maryland, US. Looking back, I can confidently say that I completed a sort of mini-MBA during the pandemic with all the courses I took.

One of the perks of living in Finland was that education was essentially free for residents (at least when I lived there). I took advantage of this, earning world-class certifications that expanded my business knowledge—an invaluable experience for someone with a PhD in engineering and a tech background. More details of this can be found on my webpage or LinkedIn.

The pandemic also gave me the rare opportunity for deep focus. This was the benefit of working in a fast-growing startup with a strong, nurturing culture. Beyond training, I immediately applied my learnings to help build Sensible 4 alongside the founders.

Whenever I learned a strategic concept, I applied it right away to our product development, bringing valuable marketing insights to top management and refining my leadership skills. The impact? We secured additional customers, and after I left, Sensible 4 received further funding from the European Investment Bank and some other additional investors—an effort in which I played a role in document preparation.

While other teams in Sensible 4 experienced high turnover, with people frequently leaving and joining, my team remained intact. Despite managing a core team of ten different nationalities, I maintained 100% talent retention throughout the turbulence of the startup phase. The secret? The future of leadership—an approach I embraced then and continue to practice today. Key elements included vulnerability (especially during the pandemic), openness, mutual growth, and transparency.

I'll share more about this in future chapters.

Soon after, the turbulence returned with Sensible 4. Looking back, I believe the main reason for this was the realization of the hype cycle in AVs. Moreover, productizing AVs is no easy task—it is not just about making a vehicle move autonomously but also about developing software that is sellable and ensures a sustainable cash conversion cycle, not only for the company but also for the customer and beyond. Securing additional funding became increasingly difficult.

Since 2020, I have also served as Secretary for the SAE Cooperative Driving Automation Committee, first under the leadership of Barbara

Wendling and later Justin McNew. Barb recently passed away. You will be missed, Barb.[4]

While at Sensible 4, I was actively involved in numerous public–private partnerships, co-creating the AV ecosystem on a global stage with our team, primarily in Europe. As a result, I contributed to two EU reports on connected and automated mobility.[5,6]

Although I worked at Sensible 4 for about 3.5 years, in a fast-growing company, one year often feels like three in a static organization. Fortunately, I have always worked in fast-growing companies, providing me the platform to launch my skills.

As the hype around AVs started to fade, I began looking for a job. Through discussions with global experts, I had a conversation with Professor Yuping He from UOIT that left a lasting impact. He said something like this, which I am paraphrasing for readability here:

Everyone is working on Level 5 autonomy (back then, the hype was all about Level 5 - there is even a company called FiveAI), but no one is bridging the gap from today's state (no autonomy) to Level 5. A lot of R&D is still required to bridge that gap, including human factors and other challenges.

That conversation sparked an epiphany. How can AVs—even at Level 4—gain public acceptance when infrastructure and public readiness are still at Level 1 or even Level 0? In many countries, even ADAS features like Adaptive Cruise Control (ACC) and Lane Keeping cannot be fully implemented and benefited due to suboptimal road conditions.

Having always networked with global experts on LinkedIn, I realized—back in early 2021—that AVs were not yet feasible for mass productization. The industry was still transitioning to Level 2 automation. Even now, in 2025, the industry is still in the midst of electrification.

---

4   https://www.barbarawendling.com/
5   https://www.enisa.europa.eu/publications/cybersecurity-stocktaking-in-the-cam
6   https://www.enisa.europa.eu/publications/recommendations-for-the-security-of-cam

That was the moment I understood: the technology itself is not the problem. What's missing is leadership—the ability to integrate technology step by step into the current ecosystem while scaling it alongside user acceptance. This requires continuous collaboration across multiple functions, including marketing, branding, and public education.

That's why transdisciplinary discussions are crucial. This realization led me to edit the book on service design and cross-functional collaboration, because I had firsthand experience of what was happening in the AV world.

In the summer of 2021, I had about ten interviews and received a couple of offers from other companies. Most of the companies I applied to focused on EV mass adoption by major automotive groups. I was close to moving to Germany to work as a Product Lead for one of the largest Tier 1 suppliers in the automotive industry when I received an offer from my manager in Sweden for my previous role—leading the EU strategy for (i) the productization of innovation commercialization from CEVT and (ii) introducing a new, then-unknown high-end EV brand called ZEEKR.

Again, luck played a role. I didn't even have a clear picture of what ZEEKR was, but I joined CEVT (which had just been acquired by ZEEKR in April 2021—I was interviewed around late May 2021) because of its growth potential. My focus before joining GEELY Group (ZEEKR) was on AV productization, but luck and the right timing gave me the opportunity to work in full-fledged EVs.

During this time, I gained extensive experience in the strategic planning for scaling AVs and software-defined innovations from the CEVT R&D team's productization into an actual product line (and not just a prototype) and importing EVs from China into the EU—one of the highest entry-barrier markets in the world.

It was also interesting because I joined ZEEKR during the same time when the ramp-up of Chinese brands' entry into the EU started to become noticeable along with other brands like XPeng, Nio, and BYD. This also provided me access to networks from similar positions and we discussed a lot of other things during this time, for example bringing products from homogenous markets and finding market fit in the fragmented markets.

During this period, I also started my EMBA at the Gothenburg School of Business, supported by my direct manager (then VP-level Head of Product Strategy) and my skip-level manager, i.e., my SVP at CEVT (my boss manager). I had actually been accepted into an MBA program in Finland but declined the offer since I was relocating to Sweden to join GEELY. Again, luck played a big role—my EMBA program in Sweden turned out to be the best formal education I have ever had (and I already had a strong educational background).

At ZEEKR in EU (formerly known to the public as CEVT before its official announcement), I initially led mobility strategic planning, focusing specifically on how to commercialize world-class R&D outputs. The result was the creation of a task force, which I led in producing the final document: the Business Memorandum for Digital Products for Next-Gen Mobility. This initiative led to my manager and me running the Digital Product Council, where we discussed the cross-functional status of this direction until a shift in priorities came.

The memorandum provided an end-to-end proposal on how to commercialize key R&D elements from CEVT and integrate them into the future CASE mobility ecosystem. It also outlined plans for integration across several GEELY's car marque brands. The potential revenue from this initiative is in the billions of USD. Of course, with strategic planning, the outcome depends on actual execution and direction. However, the output itself—a solid, scalable business case that I co-led with other departments—was a significant portfolio for me during that time. And all of this happened in less than one year after joining GEELY.

This was one of the proudest moments of my career, as the memorandum projected more than €1 billion in potential revenue if executed successfully. While its success depends on long-term execution, some of its foundational principles can already be seen in the ZEEKR x Waymo collaboration in the US.[7] I wrote approximately 70% of the document after two weeks of intensive brainstorming with my colleagues.

As ZEEKR moved into the public phase of its EU market entry, I led the initial product strategy that focused on digital, software, and ADAS-based elements for that entry. This period was quite a weird one, as we were teaching newcomers—some of whom were not from the automotive industry—about the institutionalized and tribal knowledge of the company. It provided some of my most notable exposure to big corporate dynamics (read: politics) so far in my career.

The public outcomes of my work can be seen in the introduction of the ZEEKR 001 and ZEEKR X in the EU. During the inception of ZEEKR's EU sales organization, I formulated fundamental market entry strategies for digital and software-related products, working closely with the then-CEO of ZEEKR EU, Spiros Fotinos. I had the opportunity to learn directly from top management, shadowing them in meetings and gaining insights into setting up a spinoff within a conglomerate.

Additionally, my work laid the foundation for ZEEKR Connected offerings, one of the key product pillars for ZEEKR's EU market. The initial strategy for ZEEKR EU's digital roadmap was also based on my contributions during my tenure.

Believe it or not, all of this happened in less than two years. That's how fast the pace of transformation is in the industry. I remember jokingly telling my colleagues during coffee breaks (or as the Swedes call it, "Fika"), "I'm pretty sure despite whatever our social media is saying, this pain of going fast and transformation is felt across the industry."

So yeah, that's more or less my story up until Sweden and how I joined ZEEKR. It's interesting, right? I started by moving from a PhD to a startup which has its based in Singapore with a smaller scope. Then, I transitioned to a full-stack startup for AD and automotive software management in Finland, helping the company scale. Eventually, I joined one of the largest automotive conglomerates in the world, where I led one of their high-end brands' early EU commercialization strategies for

---

7   https://youtu.be/Ew32ydR2O3c

digital and software products. This role brought me closer to the business side, all while maintaining my tech credibility.

In 2023, life threw several crises my way, one of which was big enough to force me to relocate back to my home country. Fortunately, being part of the GEELY group worked in my favor. One of GEELY's companies, PROTON, serves as its right-hand drive (RHD) arm in Southeast Asia.

I expressed my intention to join PROTON through several means, including reaching out to my previous dotted-line manager at ZEEKR, Mr. Kevin, who happened to be working at PROTON before ZEEKR. I asked him to check for vacancies. Fast forward, I got an offer to lead product export at PROTON's International Division—a scope large enough to cover product management across 18 countries.

What I did not anticipate was that PROTON was undergoing the biggest transformation in its history—a major internationalization phase. Naturally, product was at the core of this expansion. With key people leaving the division, I took on more responsibilities, expanding my role into market strategy, planning, and even regulatory topics.

As of April 2025, my role carries substantial responsibility in supporting the Head of PROTON International by leading the formulation and execution of market and product strategies. Recognizing the scope and workload involved, along with extensive cross-functional collaboration, I have also led initiatives that are unprecedented within the division, including the development of governance frameworks, new pricing models, and task forces.

In addition to these strategic contributions, I helped revamp our business processes to align with our ambition of becoming a truly global company. To that end, I led the formulation of an updated standard operating procedure (SOP) for international business within our division. This SOP outlines the end-to-end value stream for PROTON International and is derived from the global business and export strategy that I co-developed with our Head of PROTON International.

This work goes beyond strategy formulation—it establishes the framework to ensure effective execution, enabling us to operate as a global company from day one, covering market fit, product fit, logistics, and operations.

Our team, which I am leading, is also now in charge of the weekly operation governance, i.e., monitoring the performance of the international operations. In a sense, I am not only responsible for the strategy but also for the framework on how we are doing business and monitoring the operations.

It's heavy, but it's fun—almost like working in a startup within a big company inside a giant conglomerate. Some might disagree with what I said here, but I am talking based on my actual experience and contribution to a company that is undergoing transformation. I remember saying to my team members that what we are doing here is a learning curve for everyone. We are expanding very rapidly, so our learning curve for PROTON's expansion is steeper than what people in big companies experience. In big companies, they are already global, so their strength lies more in market governance. For us, we are exposed to the challenges of global expansion.

Great exposure, of course, but it's stressful and not easy. My point is that sometimes you can grow and become a bigger fish in a smaller pond. This advice comes from one of my first bosses in my very first job before the automotive industry.

In less than 1.5 years at PROTON, I've led the company's first-ever BEV export, managing the entire end-to-end task force. I also spearheaded the revamped product and market strategy for PROTON's global expansion, including the knocked down (KD) localization project for our expansion in Egypt, alongside other record-breaking internal achievements. Of course, other public portfolios will only be seen pending execution.

As of today, with the recent post-tariff announcements related to Trump's second presidency, I am focusing more and more on the impact of tariffs for international business and strategizing how to leverage this for our business success. This skill and experience, I believe, can be used and implemented across any industry.

All of this happened within just ten years of my automotive and mobility career. Fast? Seriously fast. Since 2021, I've always considered myself to have been in the same organization—GEELY. Over this time, I've faced a myriad of challenges and steep learning curves in next-gen mobility strategic management and market governance.

The key point: My career has grown rapidly, but this is no coincidence. It happened because my journey aligned with the industry's parallel growth and rapid changes. That's why I believe I have something valuable to share—my path is far from typical in this industry.

I am truly grateful for the opportunity to be exposed to global emerging markets in the Global South with PROTON, while at ZEEKR, I was more involved in the Global North (developed markets). This experience has given me a full picture of international business, complementing my tech roots. I believe such growth is possible for anyone willing to learn, stay passionate, and seize opportunities when they arise. There is no prebaked formula or recipe to follow in this scenario—we create the playbook ourselves! Of course, this doesn't mean reinventing the wheel entirely but rather carefully considering the associated risks and opportunities. This is a key message I will share in my book here.

As I wrote this, I thought to myself: Wow, young man, you've achieved something great. That's why I feel compelled to document my journey—not just as a living memory to motivate myself in the future but also to inspire others.

Many people are skeptical about my career—even I sometimes struggle to explain it due to its complexity. But after analyzing what I've done, the storyline is always the same: transformation, commercialization of emerging technologies, and leadership. The industry is evolving so rapidly that people struggle to see the full picture. It's not that I simply jump from company to company—the industry itself is changing at an incredible pace.

To the readers: if people doubt you, but you believe in your potential, chase your dream. Never let doubters define your value. The truth is that not everyone sees what you see. In my case, not everyone has been through what I have. It's normal for people to misunderstand your story.

Global Experience

That is more or less my career—fast-changing, complex, and too long to explain in detail. I will keep it high level, as the lessons learned will be expanded in separate chapters later.

One good thing about my career is that I have not just worked in tech; I have also been exposed to international experience and different functions. Let me break it down.

People often say the automotive industry is highly regional and even tribal. Each tribe has its own language, terminology, cliques, and networks—ironically, they don't really talk to each other. I've managed to work with different groups across various regions. Maybe this is something worth sharing because regardless of the environment, traversing different cultures—whether shaped by nationalism, ethnicity, or job scope—opens your mind.

Japanese Automotive

My first real international experience—or rather, the epiphany I had about working in an international landscape—happened in Japan. I was part of the Smart Mobility Research Center in Tokyo, working as a visiting researcher at the Tokyo University of Agriculture and Technology. This was because my PhD program was a joint venture between Universiti Teknologi Malaysia and Tokyo University of Agriculture and Technology via the Malaysian-Japan International Institute of Technology. My focus was on collision avoidance, but what truly shaped my perspective was the work environment.

In our Tokyo laboratory, meetings were held daily or every couple of days in 2016, where everyone presented their ongoing work. The entire team was focused on developing solutions that could be productized and integrated into vehicles, with a strong industrial vision. While the presentations were conducted in Japanese, I managed to follow along because some slides were in English. This experience taught me how to visualize data effectively and craft a storyline with supporting diagrams—an essential skill in presenting ideas.

My advisor and mentor Pongsathorn Raksincharoensak frequently attended major conferences in various countries during my time there. From him, I began to understand the importance of building connections and the impact of research beyond academia. A key realization was that a research team needs global credibility—because if no one knows who you are, why would they want to work with you? I also learned the significance of publishing in reputable journals and how this played a role in establishing influence.

Beyond research, I observed the work dynamics between bosses and employees in Japan. The culture wasn't entirely foreign to me, as I also come from the Far East—though from the southern part, Southeast Asia. These insights broadened my understanding of both professionalism and cultural nuances in an international setting.

My stint in Japan wasn't long, but I remained connected with my professor. The last time I contacted him was in early 2024. I still remember two of his quotes before I left Japan:

(i) "Don't worry. We will still meet around, in conferences every year!"

(ii) "Do we have 'another Umar' coming to join the lab as a visiting researcher after you?"

The first one stuck with me because, less than a year later, I met him again at IAVSD 2017 in Rockhampton, Australia. Then, in 2019, I crossed paths with him once more in Gothenburg, Sweden, while I was in Helsinki with Sensible 4 for IAVSD 2019. That gave me a big perspective—the whole scene is really run by people who know each other.

EU Automotive

If people really ask me, I proudly say that I grew up professionally in the EU. But to be exact—not in Western Europe, but in the North.

The Nordics. A place where vulnerability, transparency, and avant-garde, challenge-the-norm discussions on leadership often take place.

I remember that before I left Sweden, my colleague and senior Daniel Backman asked me, "What will you remember about the Nordics (Finland and Sweden) after all these years?" I was silent for a while. Then I told him—the ability to challenge people in a safe environment, knowing that differing opinions won't be met with hostility.

I have many experiences that illustrate this, but I'll highlight the most important one. When I was new in Finland, my manager—the CTO of the company during that time, Jari Saarinen—asked for my opinion, genuinely expecting me to challenge him. At that moment, I was nervous; I swallowed hard, my voice cracked slightly, but I gave my opinion. To my surprise, he accepted my constructive criticism. I was amazed.

This was in stark contrast to cultures where, in certain contexts, if your boss says the sun is rectangular, then it is rectangular. The Nordic culture encourages discussion and challenges. Of course, too much debate can make consensus difficult, but in most cases, it leads to better outcomes.

Returning to my home country at the end of 2023, I experienced reverse culture shock. This mindset—where everything can be discussed—explains why so many startups thrive in the Nordics and why Sweden is home to stable, successful corporations. I believe this culture has a profound, positive impact on society as a whole.

I will elaborate on this in the following chapters. I realize I often promise to explain more later, and if I don't, I ask for your understanding. Sometimes the scope is too broad to cover everything in one book. But rest assured, I intend to share as much insight as possible.

Returning to our topic, I also observed that the Nordics embrace the workforce diversity better—something not as common in other regions.

ASEAN Automotive

Of course, I am a native Southeast Asian—specifically part of the Austronesian people group.[8] But working in Southeast Asia is different from simply being Southeast Asian. It's almost the opposite of the EU, where work–life balance is key. Here, people push relentlessly for results. Whether working late is a good philosophy or not, it has built resilience in me during stressful times.

Additionally, working in Southeast Asia has taught me to operate with "lean resources"—making do with whatever is available. What do I mean by that?

In ASEAN, we are at the forefront of technological advancement in the region. Meanwhile, in the EU and Japan, I was just part of a large team. This pushed me to reach out through platforms like LinkedIn and ResearchGate (which I appreciate for their usability) to researchers from institutions like Chalmers and Berkeley. Some didn't reply—maybe because I wasn't wearing a "sexy" badge at the time. But being a pioneer in state-of-the-art research gave me a trailblazer mentality. I even collaborated with the then-boss of ASEAN NCAP and an Indonesian researcher to write a paper on the automotive landscape in Southeast Asia, particularly in Malaysia, Indonesia, and Singapore—key markets in the region.

That was during my PhD. Now, after years in the EU and beyond, even while working in ASEAN, people globally still take notice. I believe it's because I've built a strong personal brand while contributing real merit to our field. My point is: chase merit. Don't just chase big teams. Merit and personal skill will take you much further than simply chasing a big-name badge.

US Automotive

I have never physically worked in the US, but ironically, I have published several reports with SAE International, which is based in the US. I also published an award-winning book with SAE International. Additionally, I had the honor of meeting Jim Farley, the CEO of Ford and recipient of the 2023 Industry Leadership Award from SAE.

---

8   https://en.wikipedia.org/wiki/Austronesian_peoples

We even had dinner at the same table when I received the 2023 Young Industry Leadership Award. That's no small feat for someone who has never operated in the US.

To be honest, my experience with the US automotive scene comes from three pillars:

(i) LinkedIn connections.

(ii) SAE publications.

(iii) My role as Secretary of the SAE Cooperative Driving Automation Committee, which I've held since the end of 2020—almost five years now. Wow, time flies.

One key difference in the US automotive community compared to other regions is that it's largely dominated by middle-aged men. I've also noticed distinct groups within it—one tech-driven and the other rooted in legacy companies. The latter tends to rely on established SOPs but brings strong attention to detail, enabling them to formulate new SAE standards.

China Automotive

My experience with the Chinese automotive industry has been quite interesting. During my time at Sensible 4, we worked with a few confidential Chinese customers. From what I observed between 2018 and 2020, Chinese experts were still growing and learning, but they were learning very quickly. There was even talk back then about "The rise of Huawei"—people were beginning to notice the growth, though not fully yet.

Later, when I joined the GEELY group, I had firsthand experience working closely with Chinese colleagues, networks, and the group itself. One thing I noticed, especially in Sweden, was the dynamic synergy between the West and East. It wasn't just about understanding the Chinese side; I also saw how East and West collaborated within the same company. Gaining insight into this often unspoken dynamic has made me a better professional.

Regarding Chinese professionals, I realized they are very fast and open to collaboration—as long as you bring value to the table. For example, since 2022, I've been invited as a keynote speaker at online events organized by regional communities in China. In late 2024, I even shared the stage with key figures in Chinese automotive technology, such as Mr. Feng Shen, Executive VP of NIO and a participant in the SAE International China Vehicle Electrification and Intelligent Driving Technology Forum in Shanghai. Interestingly, he had some connection to the GEELY EU family before I joined GEELY Europe. Small world indeed. At that event, I was the only keynote speaker who didn't speak Mandarin, which shows that even without speaking the language, if you bring something valuable, the right audience will appreciate you.

But again, due to the intense competition, it is a highly demanding and challenging community. That's understandable. For me personally, it's about balancing the different perspectives and positive values I've learned from various regions and blending them into my own professional philosophy.

While my storytelling may seem disjointed, the diverse perspectives I've gained from working with different demographics in the industry have pushed me forward. Ultimately, this broad viewpoint has shaped my ability to lead PROTON's International Expansion Strategy today.

Key Achievements

I've had the privilege of working in a fast-paced environment alongside passionate individuals with a can-do attitude, which has led to some rewarding achievements. While many of them are documented online, I'd like to highlight a few here.

(i) Finnish Engineering Award 2020

This award meant a lot to me, especially since, in the early stages, our small team at Sensible 4—working on automotive software in a non-automotive country—often questioned whether we were being noticed.[9]

(ii) SAE Young Industry Leadership Award 2023

---

9   https://www.tek.fi/en/news-blogs/finnish-engineering-award-creators-self-driving-bus

At that time, I was going through turbulence in my life, facing multiple crises. But being recognized for what I had accomplished was incredibly motivating.[10] I will explain about this later.

One point that I want to bring to attention here is—yes, my career journey might not follow the conventional path of normal big careers. But it is because it is following and reacting to the demand of the market transformation direction. However, looking at the big picture, it is still following the same core of the industry, i.e., transformation. Therefore, I believe if we want to stay relevant in our career, we should really take heed of what the industry needs and where it is going. And this is what the book will discuss. And if done well, based on my experience, it will provide us with growth in skills and, of course, as bonus, getting external recognition.

Closing Thoughts

So, there you have it—the journey of my career so far. What are the lessons learned? What insights can be applied? We'll discuss them in the next chapter. Consider everything you've read so far as an introduction—to me and to this book.

Updates on July 2, 2025
The unthinkable and ironic happened. I have said that the industry is changing really fast. It is indeed very fast. When I started writing this book initially, Sensible 4 was still in a "dormant state" and inactive (the company went inactive and bankrupt a few years ago). But now, as of today, since May 2025, I am the non-operational, non-executive strategic advisor for the revamped Sensible 4, with new management and new ownership. That shows how surreal it is.

Previously, I was leading the tech and product commercialization. Now, after a few years in strategic and business leadership within GEELY group-related companies, I am back as an external thought partner in

---

10 https://www.malaymail.com/news/life/2023/06/27/malaysian-engineer-breaks-barriers-to-win-internationally-recognised-sae-young-industry-leadership-award/76586

the area of business. If I had not diversified my skills a few years back, I do not think this would have been possible.

And another point is that it just shows how fast this industry is, and if you have merit, people will always find you and bring you on board again. So unreal. Makes me only more convinced of the importance of this book.

Part II

Evolving Industry and Career Landscape

Chapter 02

The Changing Landscape of the Automotive Industry

So now, since you have reached this, you already get the idea of what my profile is and maybe a little bit about my motivation to write the book. If not, as a recap: I want to write a book that tells people, "Hey, the industry is changing. And for you—employees and young talents—here is what is changing, and here is what you need to do. And here is my story for you to benchmark, and for me to reflect as a personal reminder for the next steps."

While for the employers—"Hey, the industry is changing, and you know what? You cannot continue being complacent, because if you do not change, your competitors will kill you, and you will become the new fallen giant, taken down by emerging brands."

But what exactly has changed?

If we are discussing from the macro-level industry trends perspective, we can see trends like electrification, software-defined vehicles (SDVs), and X as a Service (XaaS). But not all of them reach mass adoption.

From what I can say, only electrification started to get traction in the past few years, but even that is not really major adoption yet.

While at ZEEKR EU and later at PROTON (both under the GEELY umbrella), I witnessed firsthand how "CASE mobility" transitioned from theory to market reality. From being involved in the project of Waymo x ZEEKR to bringing EVs to the EU from China, all of this brings stepwise CASE tech to the market. Before that, with Sensible 4 and my previous organizations, the CASE mobility elements still, at least for me, felt like the "trial" period where there was not really major production.

My work is quite interesting as I started with AVs, then moved to digital-first electrified vehicle models with ZEEKR, and now with mixed model ICE–NEV automotive business models with PROTON. In a sense, it is moving backward, but in another sense, it is me catching up with the realization of productization of the industry itself (the decline of the AV major hype in the early 2020s, going to EVs, and now starting to move upwards again from ICE to NEVs). In one perspective, I am actually following the arc of "when cutting-edge innovations can realistically be productized and monetized"—a perspective often missed when one stays purely on the bleeding edge of technology.

With the commercialization starting to kick in, and even my current company PROTON now selling and exporting NEVs, from my perspective, this actually is reshaping the value chain due to the electrification and software-defined strategies. Based on my work with SDVs, connected shared driverless ecosystems, as well as EV export planning, I can say a lot of things are changing, and it is not an overkill for me to say that every week is a learning curve and growth process as an automotive business leader and strategist nowadays.

So back to the topic—what is the change of the industry again?

But, to be honest, ever since around the end of 2022 when ChatGPT was introduced to the mass public, you can just ask Mr. C about it. In fact, there have been plenty of memes about Mr. C being everyone's best friend and personal counselor. But that is not the point.

I am writing this book *not* to sell you something that I copy-pasted from ChatGPT. I want something that is from my heart, from my mind, and basically from my own journey and lenses of life.

Changes that Happened since 2015

In Part II, I will basically tell the outside world's perspective.

I mean—what is actually happening? What are the changes? What are the challenges to moving forward? And what is the importance of being a multipotentialite in navigating the future career landscape?

To simplify, I will divide the storyline and timeline into three to show the changes. I believe we can analyze from the past ten years, as before that might be quite obsolete. So, it will be:

(i) From 2015 to before COVID-19.
(ii) Post-COVID-19 disruption and changes.
(iii) The external societal, governmental, and generational factors that might be overlooked by the leaders and society.

The world that we are living in today, in 2025, even though it looks the same on the surface, is actually *different* in reality. The same thing also happened in the automotive world. Like I mentioned in earlier chapters and my previous book,[1] we are indeed living in the period of disruption now. So what has happened since 2015?

The Foundation Shift since 2015

Maybe the most obvious changes that we are seeing are in the technology side. I will not write in detail from this perspective, since maybe millions of books and articles have been written about this. First of all, obviously, we are seeing major adoption of EVs. The adoption is quite major now that if we consider the end of 2021 or even end of 2022,

---

1 Hamid, U.Z.A., 2022, *Autonomous, Connected, Electric and Shared Vehicles: Disrupting the Automotive and Mobility Sectors*. SAE International.

most of the masses were still driving ICE vehicles. I guess the "crossing the chasm"—i.e., the transition of EVs only used by early adopters to major adoption—happened around mid-2023. This is just based on wild estimation, but at least that is what I have been feeling. I guess the rise of BYD amplified this, and it is in the same timeline as the major push for major EV adoption.

According to the World Resources Institute,[2] among the main contributors is government support (e.g., via tariff, incentives, and others, especially in China and the EU), as well as the maturity of the technology. When I was working in Sweden, I was using Polestar 2 as the company car, and during that time, the usage of EV by the public was not really still visible. But during the summer of 2023, I saw a lot of public audience start using it, en masse. Government support and maturity of technology lead to massive production, which leads to a lower cost of a full BEV. If we really think about it, the public initiatives, and the related ecosystem landscape, have been helping in pushing the technology.

However, the question is: With incentives continuing to be phased out in main developed markets, can the growth of EV still be sustainable? That is another question to show that not even the changes that we are seeing now will be promised to still exist in the future.

EVs, as I mentioned above, are part of another larger disruption and new wave of mobility called CASE mobility, which will see the ICE-laden mobility turn into connected, autonomous, shared, and electrified. At the core of all of this, there exist several foundations, including the rise of software, more advanced hardware platforms, and the change of demographics all around the world, where we are seeing more and more younger people or a digitally savvy mass population.

When I was proofreading this part of the book, I had just met a Nigerian potential partner for my current organization. One of the most interesting points he mentioned is, "70% or so of our population is less than 30 years old," which makes the country's population of that age around ~154.6 million.

---

2   https://www.wri.org/insights/countries-adopting-electric-vehicles-fastest

A similar point came up when representatives from Saudi Arabia's Ministry of Investment visited. They mentioned that about 70% of their population is under 30 years old.

I've also lived in the EU for some time, and I've noticed that much of the young talent there actually comes from outside the region. And there are a lot of reports about aging society in the developed part of the world.

And guess what is one of the most vibrant economic regions during this period of tension between China and the US on the trade topic? ASEAN. One of the advantages? Young, mostly educated in the big cities, as well as digital savvy.

So if we think about all the potential, we will see more changes in the direction of the world economy in the near future.

We are also seeing other external factors such as awareness about sustainability elements. Climate change has caused a lot of major natural disasters and rapid changes in temperature globally. This has pushed a lot of sustainability discussions among industrial experts.

The importance of sustainability has grown so much that many corporations now publish sustainability reports. These reports can influence both investor decisions and customer choices, especially for premium brands.

The same thing has happened with privacy. Since EVs collect system data, concerns about privacy and cybersecurity have grown and now influence customer conversion to some extent. I believe that if this trend continues—especially amid trade tensions between the West and East—we will see privacy and cybersecurity become key criteria for future investments, much like sustainability today.

Following the Dieselgate incidents in 2015, we have seen the tightening of regulations for the market entry of automotive products worldwide. Even for smaller countries, market-entry barriers for automakers are increasing. For example, many smaller countries and emerging economies are now embracing at least EURO 4 emission standard requirements for market entry. This shift demonstrates how the sentiment of customers, as well as regulations, have changed for the automotive industry since 2015.

We are also seeing more connectivity elements in the way human beings are living. Even before the pandemic, the baby boomer generation was still mostly using GSM and non-internet connected mobile phones. Fast forward to today, most people in major cities around the world are now connected. The same shift has happened in vehicles. We are seeing every car not only having Bluetooth but also Android Auto and Apple CarPlay. In fact, these have become highly sought-after requirements for global products, even for legacy ICE vehicle brands moving forward. If we closely examine the trends, even the market requirements for entry-level vehicles are changing, with consumers now expecting these capabilities as standard features, highlighting how connected and transformed life has become.

Additionally, we saw the rise of the digital and cloud-based services economy, which gave birth to Spotify. This, in a sense, came from the change in behavior of end users and the way they consume. It was around ten years ago when QR codes were still new. Fast forward, fintech and technology related to finance are no longer new. Nowadays, we see Wise (formerly TransferWise) and digital banks like Revolut being used by travelers everywhere. We are also seeing more and more Asian-based technology in fintech, such as Alipay and WeChat, being used widely across the world. This demonstrates how the power dynamics in technology have shifted from the West to the East. In fact, if we read between the lines, we can see that Chinese companies are not only leading in EV production but also pioneering how EVs should be marketed and sold, while many Western companies are still trying to figure things out. This shows that technology and human behavior change hand in hand, influenced by many other external factors such as geopolitics.

Finally, of course, we are also seeing the rise of AI via the ChatGPT explosion. The role of ChatGPT in human lives is now increasingly evolving, where at first it was only about "Uff ChatGPT can we trust it?" to now "using ChatGPT for everything."

From an automotive perspective, we are also seeing new entrants and business models begin to shake the traditional original equipment manufacturer (OEM) model. For example, there was discussion about direct-to-consumer (D2C) model for the car sales mode. Is it a success? That is another topic, but in a macroeconomic and bigger-picture sense, we are seeing a lot of changes in the industry. And these changes actually impact the talent landscape, employment requirements, as well as even the higher institutions models.

Some of you might be asking, why no mention of CASE mobility as a technology? Like I said, full details on CASE mobility itself can be found in my previous book here for those who are interested.[3] This book that you are reading now will touch on this topic from a different perspective.

Accelerating under Pressure since Post-COVID-19 2020+

When I'm writing this book, I am tempted to basically make it like an academic book, which means that I want to cite from academic references like my previous publications. However, the main aim of this book is for me to talk to the audience about my experience seeing the transformation in the automotive industry across different continents, countries, and types of companies—from tech garage companies to startups, scale-ups, conglomerates, pure-play companies, and legacy companies with mixed models within the same conglomerate. I believe that my perspective is quite different and may be appreciated by people who want to understand what is going on in the industry. Recently, I talked to my friend, who wanted to meet me after a while, and he said, "You have been living in a lot of countries, and you must have seen a lot of things that others may not understand or have faced." He then suggested, "Why don't you write a book to share about your experience?" and I thought, "Well, I am writing about it."

---

3   Hamid, U.Z.A., 2022, *Autonomous, Connected, Electric and Shared Vehicles: Disrupting the Automotive and Mobility Sectors*. SAE International.

If we really look closely, again, I want to emphasize this: in the companies, although older generations like baby boomers and gen X are still leading the management, the actual leaders in the middle are mostly people from the millennial generation. They have a different kind of mindset because they are the first generation that saw and lived through the rise of the internet, and they understand the transformation inside out. If you use social media, you'll see that there are many memes that say something like "the world stopped before 2019," which means that after 2020, something big happened to the world. What is that? Of course, the answer is COVID-19.

COVID-19 hit around the end of 2019 and early 2020. If we think about it, the distance from when COVID-19 first struck to when you are reading this book is almost six years (or already six years—depending on when you are reading it). If we are talking about the automotive industry as a whole and working culture in general, we actually see the rise of work from home (WFH). Technology and the COVID-19 pandemic have enabled the mass adoption of WFH. However, after COVID-19, some companies returned to the status quo. In fact, COVID-19 proved that WFH works, as many office complexes shut down during that time. But now, we are seeing the rise of working in the office and, to some extent, micromanagement again.

For me, if you are a leader and you lead like this, it is totally contradictory to the reality of the world. Some of the dotted-line staff I manage now, as well as my own staff, were born in 2002. For real. For them, 1999 is ancient classic. Before 2010? Also classic. How can we push for change if we are bringing a work philosophy that is the opposite of this?

Time flies so fast nowadays. When I talked to Sherry from SAE International about wanting to write this book, I was still around 33 or 32 years old. But now, I'm almost 35, and I think to myself, "I'm a legit adult here." This means my experience can be used as a benchmark and inspiration to the younger generation. But I am no longer the younger generation in the workforce that I was two years ago. The younger generation in the industry today is basically someone born around 1995. This means, "Are we really moving in the right direction with how we manage our talents?" Just some off-topic thoughts before we move on to the original topic now.

So we were talking about three timelines. To show how fast the changes in the industry are, especially the automotive industry—I still remember during COVID-19, most of the automakers were having problems in production. That's the news that I was reading. One of them is because of the chip crisis. And surprisingly (or not), if we really pay attention, this is because most of these chip factory manufacturing companies are located in Asia. And when COVID-19 hit this region hard, we were exposed to the vulnerability of the supply chain of the industry. I think this is also the reason why China is really strong in EVs—because of their supply chain control.

When I was in ZEEKR, in Sweden, my boss asked me to try to understand what it actually means with the new next-gen mobility, supply chain, and value stream. One of the things that I was trying to understand is basically the digital payment, the digital behavior, and all related topics such as D2C. I found a couple of companies, such as Ayden, which basically is the champion of this e-commerce-related area.

There is also this one story that I read: during the pandemic, a lot of retail clothing stores closed. But then, some of these companies studied the behavior of human beings. They realized—I mean, like one company realized—that actually, people don't need to really go to the boutique. They just want to basically have some sort of experience, a certain experience, to decide to buy.

This led to the rise of digital-driven shops during COVID-19. The changes in human behavior led us to the rise of digital business—shops and stores.

But of course, this raises the issue of sustainability and so on.

But that's another topic.

So, I believe that basically a lot of things have changed since then, and we are seeing a lot of transformation happening right in front of our eyes. What COVID-19 has done is only fast-tracking the consumer behavior changes (and to some extent human lives). The green recovery is also used by governments as a stimulus to boost the EV and tech shifts.

In a sense, COVID-19 really changed the landscape of global industry with the rise of EVs and digital economy, which is now led by an Eastern Country, China.

External Societal and Generational Factors: Cultural Shifts and the Rewiring of the Global Order

I believe if we want to really understand and comprehend the changes across all industries, we should basically not only look at the technology, like I mentioned above, but also observe the anthropology side. And this is also what I have learned in my EMBA—to observe what is not being mentioned out loud.

I believe what we are seeing today has happened because the advancement of computation is so fast that it enables a lot of the things that were only discussed in the past ten years. Just think about it: ten years ago, when I started doing my PhD, things were really new—not because they didn't exist, but because they weren't yet a reality. But now, people are not really surprised by AVs. The question has shifted to a more realistic one: How can we integrate this into our society?

So this means that not only the technology, but also how people react to and understand the technology in general, has changed.

Apart from that, we also see a lot of geopolitical events changing. For example, Trump's second presidency is expected to change the global supply chain dynamic. Early in 2025—in the first quarter of 2025 and the second quarter of 2025, when this book is being proofread by me—the Trump regime introduced some tariff changes, which I can say represent quite a major directional change toward a lot of countries. This will lead to a lot of localization, and supply chains across the automotive value stream will be changed. It will impact not only the dynamics between the US and China, but also the role of the EU in the long run.

Yes, about two—less than two—years ago, we saw that the EU imposed a lot of tariffs against the importation of EVs from foreign countries, along with many new regulations such as GDPR, which can be perceived as market-entry barriers for Chinese EVs. But the changes in geopolitics happen so fast that fast forward to now, we are seeing the Donald Trump tariff event.

So my point is that it's not only about technology but also about all of these external factors that influence simultaneously and concurrently, which require really strong leadership of companies to understand where to bring the company. The worst thing is that it needs to be balanced between strategy, tactic execution, and also talent management—among other important factors—which means that we are in the midst of a very heavy turbulence of transformation. Only the strong companies will last long.

I still remember when I was in ZEEKR, in Sweden, one of the SVPs said to me, "You know, Umar, everyone is talking about digital, digital, digital. But you know what? Less than 5% of the industry is digital."

During that time, the rise of NIO and XPeng had just started in Europe. But during that time, I was still seeing the hubris of some leadership in the industry—still denying what would happen. But now, if we really look at it—less than two years after 2022 or 2023—you will see the Chinese brands rising, not only in just the volume, but in how they are leading and really embracing the new direction of software.

Just go to China now and visit any Harmony Intelligent Mobility Alliance by Huawei, and you will understand that China is not only imagining what will happen with the next-generation mobility—they are living in something that is transforming really fast. But the reality is that Chinese brands are rising not just in volume, but also in perception and innovation.

This is the change that still cannot be comprehended by some legacy-mindset people in the Western world. And like it or not, this is the reality.

I have mentioned this before, but I would like to again mention the change and shift in the leadership dynamics. I don't have the exact number, but I can say that the people who are actually making it happen in most companies are millennial middle managers supported by gen Z. These young talents are basically pushing and challenging all norms. But, sometimes, in certain companies, they are being subtly micromanaged—or outcast—by older leaders when they have new opinions.

All of this clash in generations only causes delay in the push of the actual product and innovation to the market. And ironically, people are still saying that the new generations are not really loyal to the companies. Actually, the new generations are loyal—but they are loyal to the idea of transformation.

In addition, we are seeing, also during COVID-19, the rise of moral injury, quiet quitting, and other related mental awareness syndromes among younger-generation talent. This means that not only are they not really attracted to climbing toxic corporate politics, but they also want to do something meaningful with their life.

What is the impact in the long run on society? That's another topic. But what we are seeing is that there are changes in understanding and perception toward careers in corporates.

Apart from that, we are also seeing lesser brand loyalty, which means that no matter how good your brand was in the past, brand loyalty is no longer given, with the rise of the younger generation, which is digital-native and lives in digital experience. The mindset and willingness to pay are changing, and basically, values are the king.

That's also the reason why we are seeing the rise of BYD, Chery, and MG in the emerging and frontier markets, which is populated by young generations.

What I want to emphasize here is that with the rise of technology like TikTok since COVID-19, we are seeing a major transformation of human lives. I mean, in my previous book with SAE—if we consider it—most of the content less than three years ago is now borderline and almost quite dated because of the rapid changes in the industry. But yeah, you know, it's still a good read if you want to understand the big picture.

If you've read this far, you'll notice that the way I write this book feels like I'm trying to have a conversation with you about a million unfocused topics. But I'm simply trying to visualize how disruptive the world we are currently living in is. Later on, I will tie this back to the main objective of this book, which is to navigate through this uncertainty in the context of talent, employment, and the evolving career landscape.

Over the past decade, mobility preferences have evolved significantly. When I moved from Southeast Asia to Finland in 2018, the WHIM app in Helsinki offered a mobility-as-a-service solution. Since then, the approach to mobility-as-a-service has shifted, influenced by the pandemic and the rise of flexible solutions like ride-sharing, car-sharing, and subscription models.

Now, leasing via mobile apps is becoming more prominent, and new business models are emerging. Trends like shared driverless mobility, e-hailing, and ride-sharing were initially hyped. Some of it became mass adoption. But this hype was fueled by a growing focus on sustainability, including emission reduction, green technology, and stricter clean vehicle standards.

However, this also led to "greenwashing," where companies used sustainability claims to attract investment without delivering on them. Similarly, we've seen "autono-washing," where companies pretended to pioneer AV technology to draw in investors.

Summary

In this chapter, the evolving industry landscape and its relation to the carrier ecosystem for the next-gen industry are briefly highlighted. I divided our storytelling into three different timelines in the past decade, which are:

(i) 2015—i.e., from 2015 to before COVID-19.

(ii) Post-COVID-19 disruption and changes.

(iii) The external societal, governmental, and generational factors that might be overlooked by leaders and society.

Following the timeline, I have also shared the external factors that might influence the changes we are facing on a daily basis now due to the technological rise. The aim is to connect the technology shift with human and societal changes that we are having in the world. This shows that adaptability is not just about tech, but it is also about mindset, leadership, and understanding the market and the landscape.

Of course, we cannot have it in detail, and there are many other topics that we have not discussed. But the key here is to say that the world is changing, and the demographics of employment are also changing.

With all the trends of CASE, SDVs, electrification, regulatory shifts, sustainability, and value chain disruption—with my rapid experience, which involved me guiding ZEEKR's entry into EU markets from the context of digital and software-defined products, and how PROTON International is adapting to NEV trends (as part of GEELY)—it only taught me one major thing: that a shift from ICE to BEV, or to the more complex full CASE mobility, requires complete mindset shifts—from distribution to product design, from talent to leadership.

I believe this chapter will provide managers and professionals a macro-lens on where the industry is heading and that it will help young talent reflect on their career focus with future trends.

So, can we still expect the same leadership style during transformation, from leaders shaped in the Baby Boomer and Gen X eras, as we did 20 years ago? That's the real question.

The world, the workforce, and the rules of the game have changed. Are we ready for the change? Only the adaptable ones will survive.

Chapter | 03

Challenges and Productization in Automotive Mobility Software

This chapter aims to explore and highlight the complexity of taking automotive software-based invention from the innovation stage to scalable products and how it will impact us as employers, employees, and talents, as well as ecosystems.

Before continuing, let's take a look at the complexity of transformation aligned with my career journey, so that we have an actual real-life example to bring the weight into my argument discussion.

The Complexity of Transformation

At Sensible 4, the ambition of the company wasn't planning to develop autonomous vehicle technology. In fact, retrospectively talking about this, and in hindsight, we were trying to productize it across geographies and integrate it into the human way of life. In a sense, disrupting the masses' lives.

My role back then was not only the product owner of AV planning and control software, but the role also required aligning the R&D, product development, and business realities, which means turning the innovation of our stack into something that fits regulations, industrial customer needs, and market readiness.

In a sense, we, admittedly (at least from my point of view), underestimated the complexity. And the complexity is not on the technology or making the vehicle autonomously move from point A to point B, as that part is simple. The complexity is transforming the ecosystem itself.

I still remember one network of mine during one business discussion told me—*it is very impressive how you guys are doing this very fast, building prototypes and so on, while in big companies it is quite slow with multiple layers of decision makers*. In a sense, he is right. But it is not only about making a prototype. I learned this the hard way.

I learned firsthand that the harsh reality of software development isn't innovation alone; it's the ability to align R&D outputs with business realities. In a sense, we did move too fast at Sensible 4.

The reason why big companies are slow most of the time is because of the myriad of hurdles to change the whole value stream toward the disruption. Therefore, even though startups are fast, the most important question is how fast we can create and deploy the new technology into something that creates a sustainable business model for every actor in the value stream.

In addition, we also need to balance market demands and internal legacy debt (technical, cultural, organizational), and this is surely not easy at all.

I have basically embraced the culture of co-creation during this time where I have contributed to a number of reports by EU agencies related to this topic.

In PROTON, I also led the first export NEV roadmap, with a focus on emerging markets. This exposed me not only to developed ready markets but also to frontiers.

The key point here is: in the fast-evolving automotive landscape, building cutting-edge software or fancy AI-driven hype is no longer enough—but instead what matters is productizing it into scalable, sustainable solutions that can survive diverse markets.

That means embedding regulatory compliance from day 1, tailoring technical features to different customer behaviors, and preparing the product to meet varying levels of infrastructure readiness. It requires having a very clear vision and integrating and propagating the vision throughout every layer of your organization so that everyone moves in the right direction.

The journey was far from straightforward. Based on my work at Sensible 4, ZEEKR, and PROTON to productize our autonomous stack and EV technologies as well as connectivity for ICE vehicles, this meant balancing our technical and sales dreams with painful realities:

- Navigating strict regulatory environments, especially in higher-requirements regions like the EU.
- Localizing algorithms and EV requirements and UX for market-specific needs.
- Ensuring tech maturity aligned with business go-to-market timelines.
- Managing distributor and partner readiness, who often had different operational constraints.
- Making sure everyone understood the situation.

After seven years of productization, I have gotten more and more understanding of this topic. Bear in mind that seven years in the automotive business in a strategic position is a very highly turbulent and fast-changing period.

At ZEEKR EU, I drove connected and software commercialization strategies with the same mindset—ensuring that tech feasibility matched strategic roadmaps. Whether it was adapting connected platforms to European data privacy regulations (GDPR) or developing a NEV roadmap at PROTON aligned with emerging market conditions, the lesson remained the same:

Innovation and productization without scalability is just a prototype. And prototype only will not bring a sustainable business case.

In each case, we needed to strategically prioritize what features could truly be delivered across markets—and sometimes that meant letting go of technically ambitious ideas in favor of what would drive commercial success and sustainable growth.

The All-Around Challenges in Productizing the Automotive Mobility Software-Driven Landscape

The interest to write this book has subconsciously been there since I started in the automotive industry ten years ago. I have continuously witnessed the rapid changes in the industry—from AVs, the AV garage, and automotive software startups scaling up, to the transition to EVs within a pure-play company by a conglomerate. Now, I am moving to a mixed-model legacy company within the same conglomerate.

Throughout this journey, I have made efforts to write down my thoughts in various scattered resources, such as technical papers and presentations. One of the most significant pieces for me is an article I wrote for the SAE International Blog entitled *"Challenges of Next Generation Automotive and Mobility Innovations, Productization, and Industrialization"*.[1] In the article, I delved into the concept of productization.

To recap, according to Investopedia, productization means turning an idea, skill, process, or service into something that is sellable and marketable to the masses. When we talk about automotive software productization, we are referring to turning the idea of future

---

1 https://www.sae.org/blog/gp-umar-zakir-abdul-hamid-next-generation-mobility-productization-industrialization

technology into something that can be sold on a large scale. This term is different from industrialization, which typically refers to creating industries in a country or producing things in large quantities to generate revenue.

However, turning emerging technologies in the automotive sector into a product is not easy. In that article, I listed five key challenges, which I would like to reiterate in this chapter. I believe they remain relevant, and I will certainly add more updated insights for this new version of the chapter.

The first challenge is that software-defined products require a different mindset and way of leadership. The second challenge is attracting, retaining, and managing the right talent. The third challenge is the need for transparency across the entire organization to ensure that everyone understands the direction we're heading. Without a clear vision and strategy, we risk producing non-optimal products.

The fourth challenge pertains to legacy leadership. The issue with legacy leadership is that they are trying to lead transformation and growth phases with the mindset of leadership from a stable manufacturing era. The lack of a clear vision, strategy, and direction, coupled with gatekeeping information, leads to talent attrition. Lastly, the fifth challenge is the need for a standardized taxonomy between different organizations and functions.

These challenges remain as critical as ever in the current context of automotive innovation.

Software-Defined Products Require a Different Mindset and Way of Leadership

Before I proceed, I would like to reiterate that this book is a collection of reflections gathered from my career journey in the automotive industry. It is intended to be read from the perspective of thought leadership. Some of the observations and insights may be perceived as controversial, particularly by those who approach them from a traditional hierarchical legacy leadership mindset. However, the intention is to provoke thoughtful reflection, encourage progress, and contribute to the evolving discourse in the automotive mobility and software-driven landscape.

Actually, to lead a software-defined organization, it requires, in my opinion based on my full-fledged experience in automotive productization in different continents, a mixture of the software-originated leadership philosophy and balancing it with the strict requirements of legacy automotive. This might be surprising to you because some of you might expect me to sound too pro-software. No, my mixed experience provides me something that I believe is the right way to go—balance. A balancing act.

I still remember when I won the SAE Young Industry Leadership Award in the US in 2023, at the Henry Ford Museum in Dearborn, MI. One leader of the automotive industry of North America said that we cannot just go agile and forget that automotive is all about safety, and we need to balance between speed and quality. Of course, we can go fast, but we also need to ensure the internal sustainability of our organization, assuring that our product will not have technical debt, organizational debt, legacy debt, and cultural debt. I wrote a lot about this in my previous articles; you can just Google my name and try to read them in different outlets.

Previously, when I was at Sensible 4 and MooVita, I had partially adopted the software-originated tribal mindset, where I hated everything that came from the legacy world. But after a while, fast forward to 2025, I sort of understood why the SOP in manufacturing in the legacy world is quite strict. This is because the automotive industry has been producing a steady business blueprint for almost a century. But now, they are facing a lot of transformations from different angles simultaneously.

Some people take radical moves and fail. Some people just follow blindly what the consultants say and also fail. Pure-play companies from new organizations manage to do it, but eventually, when they become manufacturing entities, they also introduce quite stringent and strict SOPs.

So the reason why the SOPs are there is because of the need to not disrupt the plant too abruptly, because the organization is designed to manufacture. But when you are facing transformation, your organization should be designed to support the transformation. Having an obsolete organization will basically block whatever you have planned for the transformation.

On the other hand, for companies undergoing transformation, transparency is important because no one knows what is really going on, and silos need to be broken down. Walls need to be destroyed. A lot of cross-functional work needs to be done and discussed—I mean the meetings—because, at this stage, no one really knows what is going on. And to be honest with you, this is not only with technological productization. This has been ongoing even for companies facing global expansion.

I want to share a story about my current role. PROTON is a company under the GEELY umbrella, and it is actually a Malaysian brand. Now, I am basically tasked to revamp the global expansion strategy, and I am in charge of the international business strategy and global expansion planning. This means that my team that I am leading is the one that determines where we should play, how we should sell it, and what we should sell.

The thing is that because the company is quite domestic and regional and moving toward becoming a global brand, I see myself promoting and propagating throughout the company, especially in this topic, the importance of having cross-functional biweekly or bimonthly meetings, breaking down the silos. This is because we need everyone to be on the same page and basically get the buy-in of everyone.

Again, I believe once our results are out, we need to basically increase the walls again and implement a need-to-know basis. Actually, the same thing happened when we were at Sensible 4: once we grew, we started having a need-to-know basis to not confuse people.

So my point is that leadership—I mean good leadership—should know when to change their approach of leading their organization. This is what I think and believe agile really means. Agile is not a process or a holy book that has only a scrum master. Instead, for me, agile means being flexible and showing agility for the business's survival, based on internal and external limitations at that current point, to make sure the business survives.

So your philosophy should change when you are basically moving toward manufacturing by having a need-to-know basis to not disrupt the long-term sustainable harmony and traceability of the organization. This requires smart leadership.

So, take the software-origin mindset but balance it with your internal limitation. Easy? No. This requires a leader who has been exposed to a lot of experiences. And that's why this book encourages the multipotentialite features of next-generation automotive leadership.

Attracting, Retaining, and Managing the Right Talent

One thing about me that I have noticed is that I am quite good at identifying good talents. One of the reasons is that I try not to have a preconceived notion of what constitutes a good candidate. This may also be because I have frequently worked in organizations that prioritize substance over prestige. Or perhaps it is because I have developed a heightened awareness of how stereotypes can influence decision-making.

I have consistently observed that while certain individuals are able to secure roles with relative ease—despite later struggling to deliver—I was often subjected to more intensive interview scrutiny, as if applying for a CEO position. Upon joining, it became clear that privilege, rather than merit, sometimes played a larger role in hiring decisions. This dynamic extends beyond racism, encompassing ageism and classism as well.

Over time, and with more maturity, I came to realize that my approach to problem-solving often differs from others. My global, cross-continental experience, combined with my cross-functional exposure, allows me to see systems quickly and from multiple perspectives. While this can sometimes lead to misunderstandings, I now understand that this is not a weakness but a strength.

Besides that, having always worked in companies that were either just starting or undergoing transformation with lean resources, I have consistently needed to make do with whatever limited resources were available to survive. This meant adjusting the hiring strategy as well. What does this mean?

For example, when I was at Sensible 4, the budget was very limited as we began transforming into a scale-up after securing more than ~7 million USD in Series A investment. I knew that I could not afford top-notch candidates with prestigious backgrounds from Silicon Valley, but I also understood that talent in this industry is scattered.

So what I did was seriously use LinkedIn to scout for the best names by identifying young talents with aspiring motivation—people who were genuinely driven to transform the world and looking for a place to test and grow their skills. Eventually, my main team—my direct reports—consisted of 12 people from ten different nationalities. To name the nationalities, including myself, we had teammates from Malaysia, Finland, Sweden, Denmark, Hungary, Romania, Russia, India, Vietnam, and Turkey. No kidding. Even my CTO was saying, "Umar is trying to hire people from all around the world."

It's not like that, actually. The hiring happened because I'm privileged to have a systemic way of thinking. When I mapped the whole picture of our stack, it became clear that talents from Finland alone were not enough to build a strong planning and control module for the AV stack. For example, we did not only need people with software expertise—we also needed people skilled in algorithms, hardware integration, and the ability to convert R&D paper content into actual software. These are all different skill sets needed for just a small portion of the autonomous driving stack.

Imagine what's needed for full-fledged development and transformation of CASE Mobility. And for your information, talent is scattered. For example, the best automation talents usually come from Finland, but for control systems, they often come from countries like Turkey. And for software quality? Algorithms? They're all spread out.

If we do not have this mindset of embracing diversity, if we are unwilling to invest in hiring international talent and instead limit ourselves to shallow nationalism, innovation will slow down—or worse, be hindered. During a time of transformation, we must remember: no one can do it alone.

We managed to transform and refactor much of the original stack, and our team became the core foundation for Sensible 4's move toward productized software.

The same approach is currently being applied in PROTON as I started hiring. Malaysia does not yet have a large pool of global automotive business expansion strategy and product planning talents with international experience. What I do is hire based on potential and interest, and then I coach them. I provide the framework, and eventually, I serve as a catalyst for developing talents in the region.

I have always believed that good leaders beget good leaders. Leaders driving transformation should not only be focused on finding extraordinary individuals. Instead, they need to recognize that talent in the transformational era is highly sought after and that finding it is both challenging and time-consuming. Rather than seeking top-tier stars, we must identify those who are highly motivated, possess a strong learning spirit, and have the potential to develop their skills into the next generation of leaders.

This also requires the understanding that once these individuals are trained and excel in their fields, they may eventually be poached. This is where next-generation leadership becomes crucial—to retain and manage talent effectively once they are part of the organization.

Of course, if you have the luxury of hiring top talents, do so. But remember, top talent is not just about intellect. It also encompasses attitude and the ability to deliver. The most important factor is the ability to execute and understand that we are in the business of results.

In fact, I have written about this in an EE Times Europe article, where I highlighted that talent is one of the overlooked keys to future mobility productization. This insight emphasizes that one of the most critical factors when discussing talent requirements for future mobility and related fields is not the size of the resume, but its composition. The individuals we hire will define the culture of the company, especially in the early stages of transformation.[2]

Ensuring Transparency across the Organization to Provide a Clear Vision and Strategy

Recently, I browsed through my LinkedIn and noticed one article that really attracted my attention: *"Stop Working for Loser Strategist,"* or something like this. The title really captured my eyes because I have seen many times when the delay of innovation, business growth, expansion, or changes happens just because one person does not understand what it is all about. And this is after rounds and rounds and rounds of discussion between the other leaders.

Who is that one person? Often, it is someone holding a critical decision-making role.

So sometimes, no matter how good our experience and strategic outputs, if the final decision maker approaches the situation with a purely operational mindset, all of our plans will not work.

I still remember when I was talking to one of my dotted line leaders in Sweden. He said something like this: you know, strategy—a good one—is basically making sure everyone from floor to the top understands what is going on.

For example, if let's say we put our strategy as making sure that our car is safe, or maybe simpler than that—we produce safe cars—this means that everyone from the floor to the top understands their task. Their individual task is to make sure that our car is as safe as possible. For example, people from the floor will basically make the best seat belts that they can, the best sensors that they can, and the best algorithms that they can, to make sure that the vehicle we deploy is the safest car.

That is a good strategy. However, this is in an ideal situation.

In the case of transformation, sometimes the change happens very fast and no one knows what is going on. And here comes the leader's

---

2 https://www.eetimes.eu/talent-overlooked-key-to-aces-mobility-productization/

role—to basically decipher this and have frequent town halls to make sure that everyone understands where we are going.

Here comes the importance of having humble, transparent, and open leadership that can take criticism and also critical constructive feedback, because no one knows exactly what is going on, and our employees want to help us be better.

Hubris of leadership is not what we are looking for when we are going through transformation and a new era. But of course, it needs to be balanced with business importance because by the end of the day, you cannot satisfy everyone. Like I said, the balancing act.

In addition, making sure that innovation work is done is not the task of one particular department only—it is teamwork that should be done by the whole organization.

It will not be possible if only the business team, technical team, or manufacturing team wants to drive the change.

So, a clear vision from the top is expected. Therefore, it is very important to communicate this regularly and take feedback from all directions.

Because unclear strategy, unclear vision, and uncommunicated vision can cause a lot of waste—not only financially, but it will also cause a lot of burnout across the organization.

This means that transformation requires a different kind of leadership. And it requires a different kind of decision maker compared to a stable period.

In other words, if you want growth and transformation, hire a certain kind of leader.

If you want stability and quick action to save a sinking ship, hire a different kind of decision maker.

And if you want to scale sustainably for the long run, again, a different kind of leader is needed.

Of course, you might find one person who can do all three—but in reality, it is rare. Legacy leadership often struggles with transformation because their mindset is rooted in the stable manufacturing era—resulting in lack of vision, unclear strategy, weak direction, and ultimately, talent attrition.

Actually, one of the chapters in one of my previous books, which I co-edited with my friend Mari, talks about this: during transformation, what is needed is circular communication, not just top-down or bottom-up.[3]

The Need for Standardized Taxonomy between Different Organizations and Functions

When I was drafting this part of this book proposal—the first draft—I basically was thinking to write about the idea of talking about different types of taxonomy between different departments, continents, or even organizations when we are talking about AVs. That was the original notion of this chapter.

It's quite funny—or is that even the right word—that fast forward a couple of years, when I am actually writing this part of the book, I now have a larger frame of definition of what this actually means.

Now, my role is about devising the international business strategy, which means our market strategy, global product strategy, and also other localization strategies. Basically, doing the strategic planning and leading the export strategy. I am also responsible for the governance of weekly market monitoring as of now. So, I have a clearer picture of the whole end-to-end automotive international business of our company and the industry.

As such, when referring to this subsection title again, the concept that I am writing now is also changing. Larger scope, i.e., company transformation.

For transformation to happen, we need to make sure that everyone in the whole organization understands exactly where we are going—with

---

3 Walter, S., 2023. Designing Human and Artificial Intelligence Interactions in Industry X. In *Service Design for Emerging Technologies Product Development: Bridging the Interdisciplinary Knowledge Gap*, pp. 207–232. Cham: Springer International Publishing.

the right terminology. Let's say if the business strategy team understands that when we are talking about global expansion, we will go for developed markets, then we also need to make sure that everyone in the organization understands what this means.

Of course, we can play with ambiguity, but this is the role of top leaderships—to basically always bridge the communication between different divisions so that the silos are broken down. If not, people will only attend meetings like headless chickens, and basically no proper chain of command exist. And burnout spreads like wildfire. And talents leave the company rapidly.

Looking back at the original reason why I wrote this subsection title, it's because sometimes arguments between different experts, professionals, or employees can happen because of the non-uniform jargon and taxonomy that are being used.

For example, if we talk about autonomous mobility and self-driving, we have seen a lot of times that the marketing and communication team propagates some vehicles as having self-driving capabilities. But instead, the engineering team might disagree and say that this is not self-driving, asking "What level of automation are we talking about here? Level 2 or Level 4? Or Level 5?"

This might seem simple, but having clear definitions, especially when we are talking about crucial taxonomy and terminology across businesses, is important. Therefore, all organizations facing transformation should have a sort of an internal dictionary that is owned and edited by key stakeholders so that everyone understands the same terminology when referring to these crucial key points that will influence decisions.

This especially happens if the transformation happens too fast and the team is decentralized, talking about emerging technologies and product positioning.

Luckily, a lot of effort has been made by SAE standardization committees to address this. For example, we have SAE J3016, which examines the different levels of driving automation. SAE J3163 discusses the taxonomy of on-demand and shared mobility, ground mobility, aviation, and marine, while SAE J3216 integrates the taxonomy and definition for terms related to cooperative driving automation for on-road motor vehicles.

This is also the reason why, since the end of 2020 until now, I have been basically the secretary for the SAE Cooperative Driving Automation Committee, because I want to bridge my understanding of the taxonomy and understanding used between different continents.

As far as I understand—having the experience working and/or researching in Japan, EU, Southeast Asia, collaborating with professionals from the US, and working under the umbrella of China-owned conglomerate—I realized that people in different organizations and continents are using different terminology even though everyone is using English.

This exposure of different experiences made me realize this. And that's why I believe the importance of being a multipotentialite to stay relevant in the next-gen industry transformation as a talent.

Please note that some of the contents in this subsection are taken from my article in an SAE Blog in 2022 that can be read here: https://www.sae.org/blog/gp-umar-zakir-abdul-hamid-next-generation-mobility-productization-industrialization.

Conclusions

In this chapter, we discussed the challenges and product decisions in the automotive sector related to the next generation of emerging technologies. This is the continuation of the previous chapter that talked about the evolving landscape of the automotive industry.

As the automotive world becomes increasingly software-defined, and as we are seeing a lot of geopolitical changes and new household names emerge, the ability to bridge innovation with execution will define the next generation of leaders.

The key takeaway for aspiring professionals, employees, and also employers for these sections is that it is very important for us to be flexible and nimble when we are talking about getting our skills expanded because the changes in the industry are very complex.

For future leaders of the industry, it is crucial to understand that success is not only about having the best technology, but it is also about solving and delivering what the real market needs, and this should be done sustainably.

In addition, for managers, recruiters, and organizations that are looking for transformation, we should look for candidates who not only think about the problem as a specialist, but also think about scalability, localization, and user adoption.

This means that we need to open up our mind in regard to who we think is the best talent.

In the next chapter, this book will explore how the career landscape itself, in reference to the automotive industry transformation, is evolving.

By embracing a diverse skill set—in other words, by being a multipotentialite—it becomes important to be exposed to a lot of things and also to be willing to put in a little bit of extra effort.

This means having someone with diverse interests and capabilities.

Therefore, having this kind of people is no longer optional but necessary in our organization, because the transformation happens really fast and we need someone who has different kinds of perspectives from different angles to understand what is going on and to lead our organization moving forward.

Chapter 04

Evolving Career Landscape and Multipotentiality in Automotive Mobility

Traditionally, it takes maybe five years for someone to become a project manager in a manufacturing legacy company. Then, it will take another three to four years for someone to climb to senior manager, followed by another five years. In total, maybe it will take about 20 years to become a vice-president.

However, with the changes that are happening so fast in the industry, is this still logical? If we think about it, the mass production of EVs only happened less than a decade ago. Is someone who has seven years of experience in pure-play EV companies, with exposure to state-of-the-art software-defined automotive business and leading projects with a proven portfolio, less than someone who has been working 20 years manufacturing ICE in legacy companies?

My point is that the career landscape in automotive mobility is changing, and therefore, the requirements are also changing when we are selecting the leaders. The shift from traditional linear careers to multidimensional roles does not happen overnight, but it is not yet fully understood or seen because the transformation happened so fast.

As employers, if we are looking for someone to lead our organization in a new direction, we should look for someone who is a mixture of different dimensions and trust their decisions without micromanaging them. The multidimensional perspective is important, especially when an organization is going through transformation.

This also means that organizations should prepare themselves to handle this kind of talent and prepare a dedicated framework and special structure to support the new career path for this kind of talent, which will allow their multidimensional capabilities to thrive well in the organization.

I still remember when I went back to Malaysia. One recruiter said, "Your talent is quite rare, even in Southeast Asia, because you have worked in a developed market in electric vehicles and you know the front line of transformation in the industry. Therefore, if I want to give you a position, I will give you a position that can influence the whole organization, because if not, you will be working and executing someone's direction that might be obsolete, and it will only kill your talent."

This is the same situation that I believe is happening to many emerging talents in our industry. That is why companies nowadays have this sort of skunkworks organization that allows top talents to operate and show their potential without many constraints. But of course, this requires financial luxury.

And to the talent itself, sometimes it is better to work in an organization that will allow you to thrive well with your skills and trust you, rather than work in big companies that basically micromanage you and disrespect your skills. If not, you will basically have imposter syndrome and basically doubt your capabilities and lower your ceilings.

I still remember one quote from one of my dotted-line bosses when I started my career: "Sometimes if you want to be a bigger fish, you need a smaller pond."

It depends on how you want to understand it, but for me, it shows that sometimes if you want to learn and grow, you need to work in a more supportive, smaller environment rather than working with people who are doing politics every day.

For me personally, my career has always been about taking one step at a time and grabbing the next opportunity that I see. If you are coming from a control engineering background, there is this control system called model predictive control where, to put it in layman's terms, it is basically the prediction of the next step and the next action, depending on certain information within a finite horizon. This means that's also what I have been doing with my career.

I join a company, I work and take as much learning as possible, and then after that, I basically decide my action based on the current situation at that current instant. Sometimes opportunities just come and give you the opportunity. But to reiterate, I never really switched companies without any reason, without any loyalty to my career trajectory or career storyline.

It has always been about transformation—never about money or status. In fact, there were times I even reduced my salary to keep growing. I never intended to remain just an engineer, because growth has always been the central focus guiding my decisions. Over time, that mindset has shaped me into a global business strategist and expansion leader. In today's automotive world, futureproofing your career is no longer optional—it's essential.

Sometimes it just happens. But to make it happen, and for that opportunity to come and for people to notice us, we need to expose ourselves and grab as much as possible when the learning curve happens.

Understanding Multipotentiality

The automotive, mobility, and transportation industry is no longer shaped solely by specialists or single-domain skill experts. To lead in the next generation of mobility and the automotive industry, leaders must possess a combination of strategic perspective, technology knowledge, market-facing, and customer understanding, as well as storytelling capabilities. These skills are essential to gain internal buy-in from stakeholders and navigate the increasingly complex ecosystem of the automotive industry.

For example, in my case, especially since my days at Sensible 4, I often found myself making presentations to gain buy-in from stakeholders. If you want to sell a product, communicate its value to stakeholders, secure investment, or gain top management's approval for a project, presenting your ideas concisely is vital. People simply don't have the time to engage with long-winded explanations. So, how can you explain complex problems with concise, precise, sharp, and firm words?

This is where the ability to plan and explain from different perspectives, as I mentioned earlier, becomes critical. And this is where multipotentiality is important, having the ability to understand and connect different domains to address challenges effectively.

These skills are indispensable for leaders in today's automotive and mobility industries, where success lies in blending technical expertise with strategic insight and effective communication. As I mentioned earlier, my profile is truly diverse. It may take a couple of hours to fully explain what I have been doing over the past decade. However, to simplify, in my context, multipotentiality means having a combination of strategic thinking, technology experience, and global market expansion.

All of these integrations enable me to operate at the intersection of product, business, and market realities. This allows me to comfortably switch between contexts, using different terminologies and language when interacting with various departments. Since joining GEELY, I've found myself leading cross-functional conversations and translating what one department is saying to another, especially during heated discussions. Sometimes, this involves translating what the Chinese teams are explaining to the European teams or what the Southeast Asian teams are conveying to the Chinese teams. And of course, during these discussions, a few light-hearted jokes are always welcome to help reduce the tension.

I'm grateful that I can understand different perspectives and the reasoning behind them, thanks to my diverse experiences and exposure, as well as my combination of various educational backgrounds.

Multipotentiality, by definition, refers to the ability to excel in multiple fields and integrate them into unique, valuable perspectives. It is the ability to traverse between different departments, divisions, company cultures, and continents, while seamlessly conveying complex systems into understandable, actionable items.

In the modern automotive mobility space—where electrification, digitalization, regulation, and globalization collide—multipotentialites are becoming crucial. They serve as bridges between silos, connecting different departments while everyone remains isolated. They can see things as a system, which enables fast decision-making.

The rise of multipotentialites is not accidental. It reflects the industry's urgent need for talent who can navigate between engineering complexity, market needs, and corporate strategy. This ability is increasingly critical as technology reshapes mobility faster than legacy structures can adapt.

However, unfortunately, these talents often do not receive the support they need. In many cases, they are seen merely as tools, despite doing the heavy lifting. As a result, valuable talent leaves companies, leaving organizations stranded in their transformation efforts. But for talents to become multipotentialites, they must expose themselves to as many learning curves as possible, even if it means learning lateral skills rather than seeking immediate promotion. I remember when I was doing my EMBA, one of the lecturers emphasized that what we need in the future are T-shaped professionals—individuals with deep expertise in one area, but who can also control and lead conversations across a wide range of topics.

Personal Journey: How Diverse Skills Shaped My Career

My path was never linear—and that became my greatest strength. When I was younger, I did not realize this. However, later on, when I began writing this book, I realized that my so-called struggle is what has given me strength.

The combination of an EMBA from the Nordics, a PhD joint venture between Malaysia and Japan, and multi-country experiences with both startups and the corporate world equipped me to transcend traditional engineering boundaries, stepping confidently into strategic leadership roles. Am I proud? To be honest, if you ask me, I would say I am still learning. And I want to grow further. Because, by growing, you stay relevant—and more opportunities come.

From the perspective of experience and portfolio, this diversity of expertise enabled me to:

- Contribute to the €20M+ collective funding and revenues at Sensible 4 (investments, public projects, customer projects, and winning some awards).
- Work with the team on ZEEKR's first entry into the EU from foundational strategic perspectives, as well as influencing the long-term direction of ZEEKR Tech EU's innovation roadmap (previously called CEVT).
- Navigate the complex internal ecosystem that enabled the first foundational strategic work for PROTON's first NEV export in early 2025.

Each achievement was not about being the best engineer or the best strategist—it was about integrating multiple perspectives to drive real-world results. This is the identity I am becoming more comfortable with day by day in my career. I believe that if we can have more people with cross-functional or multipotentialite skills, we can expedite a lot of things within companies.

The Role of Multipotentiality in Automotive Mobility

A lot of cases confirm that success of career growth in today's mobility world demands more than depth—it demands breadth and integration.

In automotive mobility, multipotentiality is materialized when multiple skill sets converge to address the industry's multifaceted challenges. Whether it's electrification or connected mobility, the industry's most pressing issues require leaders who can step into a role beyond their job title—combining tech skills, business strategy, and global market insight to shape the future. And this will basically be useful because such leaders will basically be able to lead the direction and steer the ship with just good enough force to prevent crash or even understeer when crisis happens during fast transformation.

At Sensible 4, for example, I was not just developing AD technology—I was also basically building the company considering a significant chunk of the engineers in the organization during that time reported to me, as well as designing the architecture of the AD stack as well as cross-functional with other departments for due diligence related to the investments. This integration of skills was critical to help us build the required solutions while assuring the livability of the company itself. But of course on paper, I was officially called the Product Owner (in some companies might be called Product Manager). But again, is it a right term to actually portray what I delivered?

Similarly, at ZEEKR EU, I relied on my diverse experience to blend digital innovation with strategic go-to-market planning. From product-market fit to regulatory compliance and consumer expectations, having a broad skill set allowed me to bridge gaps between R&D, commercial teams, and external stakeholders, ensuring concrete strategic planning outputs for both ZEEKR product lineups as well as innovation commercialization for CEVT during that time. For example, the ability to traverse between different functions pushed me as the internal subject-matter expert (SME) for the GDPR despite not having such experience before, allowing me to work closely with the legal department during that time, devising plans for ZEEKR EU entry from China.

Again, of course, the success of all of these planning and strategies depends on the execution. But that is another story. That's why I said: don't limit your title or identity to what a company calls you. In fast-growing companies, you are, in most conditions, undertitled—and maybe also underpaid. But the most important thing is to keep building your resume, build your CVs, and build the right professional branding for your identity.

Requirements for a Future-Proof Team

As the industry evolves, so too must our approach and perspective to talent. The multipotentiality mindset is not just important for individuals—it's vital for how we build teams. Roles in automotive mobility no longer fit into rigid boxes. An engineer today must understand regulatory frameworks and business requirements, an analyst needs to speak the language of technology and marketing, and a strategist must also be fluent in execution.

In my current organization, although there is general support from across the company, when it comes to international business, it is only a small group of dedicated team members who are consistently working to put PROTON's name on the global map. As such, for my team member itself, for the market strategy, as we don't have the luxury to hire a lot of individuals, the lead of market planning should have a combination of skills in international business, project management, budget request financing case study feasibility, cost reduction, market analysis and insights, customer study, as well as global product planning.

But to be honest, having too many individuals who are too specialized and too rigid in giving effort to things that are outside of their job scope will only slow the progress of transformation. Multipotentialite talents are a must for companies undergoing transformation, as they simplify the task and provide speed of execution, as well as acting as in-depth T-shaped specialists.

The *downside*? Leadership really needs to take care of the talent, especially the young ones, so that they won't burn out and be poached by other companies.

Speaking of this, when I am hiring, I've always looked for hybrid talent—individuals who bring cross-disciplinary skills to the table, rather than just deep functional expertise. This enables teams to be more agile and adaptable, capable of solving complex problems that arise in international market expansion or digital productization. The future of mobility demands it, and so does the future of leadership.

Criteria to Nurture Your Multipotentiality for Future Relevance

This will be discussed in detail in the following chapters, but briefly, for young professionals who aspire to stay relevant in industries undergoing rapid transformation, nurturing your multipotentiality is the key. Multipotentiality is about embracing lifelong learning. Always learn something that will benefit your career storyline trajectory and try to reduce your energy doing something leading to no results. In the fast-transforming automotive industry, to stay relevant and attractive in the job market means continually acquiring new knowledge and getting the right environment to grow.

This requires you to diversify your knowledge. Always be not afraid or limit yourself to the discussions of one field. Always try to see the big picture of any decisions you are making, i.e., the macroeconomics, the impact not only to your companies but also to your parent group as well as the automotive industry as a whole. Having said that, knowing other functions like marketing, communication, or engineering, depending on your job area, will help improve your decision-making and make you more attractive in the job market.

For productization of emerging technologies, instead of just "building," aspiring professionals should also have this big picture, i.e., strategic mindset. This means try to understand not only how the product should be built, but why. By having this mindset, we will be able to scale the product, being nimble with all the opportunities, and scale them across diverse regions and regulations. Especially for international business, it is important to understand the big picture when devising the strategy—for example, why people in those countries even need our product first.

In summary, to be a truly attractive talent in the job market, multipotentiality is a must, which will basically provide you with technical depth with strategic breadth across all functions and organizations. By having these features, we can become the leader who can navigate ambiguity with confidence and calmly lead global success across borders.

Conclusion

The future of automotive mobility is not about rigid roles; it's about embracing diverse skill sets and nonlinear career paths. In fact, I believe that more jobs in the current markets will become obsolete, and we will start seeing more and more new titles and jobs, especially with the arrival of younger talents.

Since the most of the current legacy organizations are not built for transformation but basically to maintain the status quo, professionals who can adapt, integrate multiple disciplines, and bridge gaps between different fields will be the ones shaping the future of mobility. As the automotive industry moves toward increasingly SDVs, electrification, and global market expansion, the ability to think across boundaries becomes more essential than ever.

For aspiring professionals, embracing multipotentiality allows you to stay relevant and innovative as the landscape continues to evolve. I believe that this chapter will empower aspiring professionals to embrace multipotentiality, which will position them to thrive in the future of automotive mobility.

For recruiters and organizations, this insight is invaluable and is a must to be implemented. Instead of focusing on candidates with a single skill set, recognize the tremendous value of hiring individuals who bring hybrid skills that are capable of navigating the complexities of modern mobility. These are the people who will thrive in the shifting sands of tomorrow's industry.

In addition, this requires us to open our minds and actually look for talents instead of just screening the physical appearance. Finally, the chapter encourages a broader vision—hiring individuals who embody the hybrid skills and flexibility necessary to lead in a fast-changing world.

Part III

Lessons Learned and Reflections

Chapter 05

A Decade in Automotive Mobility: The Role of Multipotentiality

Reflecting on my decade-long journey in the automotive mobility software industry, it's clear that multipotentiality has been a defining factor in my growth in the automotive industry. From starting R&D in ADAS and AVs, then moving forward to technical contributor to leading startup productization and helping them move to the scale-up stage, to leading global expansion efforts for GEELY groups via ZEEKR in the EU and PROTON in ASEAN and Global South, every transition in my career has highlighted the importance of wearing multiple hats, building cross-functional expertise, and evolving with the industry's fast pace.

Key Transitions, Milestones, and Turning Points

From Tech Contributor to Team Lead

I will only highlight my automotive career starting from Finland here, because before that, I was in the R&D phase, and in MooVita, it was a garage startup phase. Despite that, actually, that prior experience before Finland provided me with the can-do and grit attitude, as well as the underdog mentality that can build something with lean resources. Not a bad thing.

My significant automotive milestone began at Sensible 4 in Finland, where I immersed myself in deep-tech. From tech contributor to team lead and early strategic role, that's my storyline in Finland. Sensible 4 was also, during that time, co-owned by GIM Robotics, a spinoff from a very household name of a Finnish robotics team of researchers from originally Aalto University (previously with a different name). This exposed me to the end-to-end deep-tech landscape of Finnish technology—think engineers that actually use Linux with four screens flipped horizontally and vertically, wear hoodies 24/7, and eat all of these like their breakfast.

As well as working with Sensible 4, I was also involved in some projects related to the AD stack for robotics applications with GIM Robotics, where discussions also involved how to streamline and communalize the stack for robotics. For example, areas that we discussed basically included surveillance, marine applications, satellite, and, even to some extent, potential of defense. But all of this was me working as a subcon internally to GIM for a few hours daily since they were our sister company. However, the exposure and agility introduced me to a legitimate robotics deep-tech ecosystem, and not only AD.

Of course, during this time, because the work–life balance is quite good in Finland, I managed to spend more time with my kids and also pushed myself to write a lot of books and articles about what I had learned at the office. During this time, I started to act like an intrapreneur and talked to my networks about the industry, continuously posting on LinkedIn and getting some new connections. Eventually, my name started to gain some traction, and I managed to output a couple of credible publications and a book. Quite impressive when I think about it, considering I had not even reached 30 years old and was a non-EU citizen. And as you can see, it's not only about what I did at work, but what I did outside of work.

I have also been exposed to EU public projects. So despite it feeling like doing everything at once, I learned a couple of key things here: I learned the grit of the startup business, and I also learned what not to do as a business leader, as well as the importance of prioritization for long-term business strategy.

After investment, I was given an official task as a product owner and engineering team manager. I led the development of an AV stack, overseeing a team of 12 engineers (direct reports, not including the dotted line). But not only that—we had just secured our Series A investment, and for the first time, we began to build the company like a corporate entity. I was the main coach of the technology to the new staff, including sales teams, and I was a core part of forming the culture (indirectly). I attended a lot of cross-functional meetings and implemented the leadership style that I believed should be done.

Growing the company from 20 to 100 people in a short time created an earthquake—an earthquake that management of companies usually underestimates. That's why I believe another lesson learned here is that if you want to transform, you must also include headcount increase as a stepwise plan in your tech roadmap as well as a realistic timeline for integration into the organization of these new talents and when they can start delivering. This is one of the keys that I believe only a few people can understand, but is necessary. Too late if you keep doing slides and only think about hiring at the end of the pipeline. HR, legal, and accounting should co-create and come into the picture as soon as possible.

Consequently, with all these next-gen leadership philosophies I embraced—transparency, vulnerability, and continuous feedback with clear expectation—I managed to retain our 12-member team as 12 members up until the day I resigned, even while other teams saw their members resign every now and then during the very turbulent period of company growth.

This role taught me the importance of blending technical depth with market needs. The results might not have always been there because my authority was not very strong in the context of business ownership during that time, but I know what leads to it not being really successful. Again, for me, understanding who your actual customers are, building what they actually want to pay and use for sustainable long-term market growth, and communalizing as much as possible are among the key lessons.

At Sensible 4, there were many achievements, but to highlight a few: I was one of the tech leaders who successfully supported the expansion of the engineering team by 300%. During this time, we also launched a complex autonomous driving software stack that integrated motion control, decision-making, and risk assessment.

It's worth noting that Sensible 4 initially started as a positioning and mapping company. I took the initiative to introduce and develop automotive-level planning and control algorithm stacks. It wasn't perfect, but it was solid and reliable enough to attract top talent and secure significant industry customers.

With lean resources and limitations, we managed to deliver a lot of prototyping to corporate companies as well as public projects in cities in Norway, Japan, several European countries, and Finland, with big names too collaborating with us in the public projects. And this all started with less than ten people when I joined the company. Of course, I believe I can take credit and document these here. Even though it was roughly ~3.5 years there, it felt more like 10 years of experience considering the challenges we endured.

Sensible 4 was then declared bankrupt in 2023, but the experience that I gained there—pure deep-tech startup rise and fall—is something not everyone has gone through. Very valuable. I can even write another book about it—"Startup Playbook: What Should Be Done and What Should Not Be Done." Maybe later though. If you are interested to organize trainings for this topic, you know where to find me. Full details about my story can be found on my LinkedIn page. I always wrote there in case I remember something that could be learned by other people.

As I have mentioned in the earlier part of this book, as of July 8, 2025, I am now back with Sensible 4 in the role of an external thought partner, serving as their strategic advisor under the new management team, which is currently working to revamp and pivot the company. It is indeed a very interesting journey.

Pivot to Strategic Innovation (ZEEKR)

My time at ZEEKR marked one of the most pivotal shifts in my career—from deep-tech contributions to driving strategic innovation at a corporation, with billion-euro scale potential.

I first joined ZEEKR R&D Europe, operating under CEVT, as the Lead of Strategic Mobility Planning. My focus was on the commercialization of future mobility innovations within the GEELY ecosystem. This role wasn't just about product planning—it was about influencing corporate business directions and shaping the early foundation of ZEEKR's future in Europe. I was immersed in building strategies that could potentially unlock billion-euro business models, particularly for next-generation mobility.

Even until today, I believe that B2B commercialization models are key if we truly want to unleash the full potential of CASE mobility. (More on this—feel free to contact me separately.)

In that first role, my scope included:

- Leading strategic planning for CEVT's "Future Mobility" product portfolio at ZEEKR.
- Preparing industrial market deep dives and defining product visions—critical elements that anchored future product definitions.
- Co-authoring business, investment, and offering memoranda to align technical innovation with business reality.
- Driving cross-disciplinary collaboration across departments through the Future Mobility Portfolios Steering Committee, attended monthly by CxOs, SVPs, and VPs.
- Deep-diving into global and EU-specific automotive and mobility regulations to ensure strategic compliance, particularly in digital and autonomous sectors.
- Acting as an SME on data privacy and regulatory compliance frameworks such as RoPA and PIA.

It was a crash course in thinking like a global business corporate strategist—connecting technical possibilities to market realities, governance structures, investment cases, and board-level decision-making.

Promotion to Senior Lead Strategist and Product Planning (Digital, Mobility, ADAS, and Software Product).

Less than a year later, I was promoted to a more externally facing strategic role, at the center of ZEEKR's move to enter the European market.

Here, my focus shifted to driving real-world market entry—not just preparing the internal strategies but executing them. CEVT remained the legal entity during the initial phases before ZEEKR was formally launched to the public.

In this elevated role, my responsibilities expanded to:

- Driving automotive and mobility digitalization strategy across Europe.
- Shaping product strategies for ADAS, digital services, connected vehicle ecosystems, and omnichannel consumer experiences.
- Leading long-term strategic planning for the SAE L2–L4 AD transition, applying frameworks like Porter's Five Forces to competitive analysis.
- Designing the omnichannel retail and charging ecosystem journey for ZEEKR's European customers—including strategic input into zeekr.eu digital platforms.
- Managing cross-continental collaboration between EU and China stakeholders for product operations, digital lifecycle management, cybersecurity, sustainability, and operational excellence.
- Partnering with pricing experts to formulate competitive market entry strategies, recurring revenue models, and software commercialization plans.
- Driving change management initiatives tied to digital transformation, internal talent development, and cultural alignment between East and West.

In simple terms: I was not just theorizing the next wave of mobility—I was helping bring it into reality.

If you visit ZEEKR Connected today (https://www.zeekr.eu/en-se/connected), you'll see the outputs of those efforts—a seamless ecosystem where vehicles, software, and consumer experience converge.

It wasn't always easy. The European market is fragmented, highly regulated, and fiercely competitive. But that complexity pushed me to become sharper, faster, and more strategic, balancing startup energy with corporate discipline.

Global Expansion Leadership (PROTON)

When I came back to Malaysia, it wasn't something I had originally planned. I had a major private situation at home that required me to return—no choice. Even my bosses in ZEEKR EU and my EMBA friends supported my intention. At that time, I wasn't looking to jump companies randomly. I've always believed in building long-term compounding value, not just chasing roles.

Luckily, I still had a strong network within the GEELY Group. My former dotted-line leader—who had been at PROTON before and was at that time at ZEEKR HQ in China—knew me from my time at ZEEKR Europe HQ. I reached out to him first, asking if there were opportunities. At the same time, I contacted people at DRB-HICOM, PROTON's co-parent company. Even though a few external recruiters also approached me, I made a very deliberate choice: I only wanted to stay within the GEELY family. I believed in the group's long-term vision and wanted continuity, not just a career "move." I believe it will provide me with international business expansion skills.

That's how I ended up joining PROTON. It turned out to be a very unique, almost serendipitous move. I initially came in to lead the global product export program to 18+ countries. It had a startup vibe—very dynamic—but the impact was at an international scale. Little did I know at the time that PROTON was about to embark on its most ambitious global expansion and electrification transformation ever.

Of course, with any major transformation, there's always turbulence. Many key people were leaving (which, to me, is normal during a transformation era). This situation unintentionally positioned me as the de facto no. 2 of the International Division (with no official promotion—but for me, the growth itself will give me a lot of money in the long run, and that's what I have always been doing throughout my career—take the jump, leap responsibility, and grow your skills. I know this is controversial, but sometimes you need to know what is your priority). I became the internal connector across departments—commercial, product, operations—and eventually, my scope expanded from leading product export to essentially driving the end-to-end global expansion strategic program.

Today, less than 1.5 years later, my role is heading the International Business Strategy, Global Product, and Market Expansion. I'm driving PROTON's global expansion strategy, market positioning, and product strategy across 18+ countries—leading market entry, business growth, and electrification exports (both ICE and NEV models with connectivity).

Some key milestones that I obtained in a very short period (like I said, it's a startup within a big group within a big global conglomerate):

- Market Entry and Business Strategy: I architected PROTON's international growth long-term strategy, introducing frameworks needed, producing CXO-level strategies for ASEAN, Indian Subcontinent, and African markets. Our export volumes are up year-over-year, from 3,636 units in 2023 to 4,765 in 2024, with sustainable momentum. I know this is small, but I believe if executed well, we can be a legit global brand soon. It's just a matter of execution now (which is not my job alone).

- Global Expansion and Corporate Strategy: I defined the international growth roadmap, secured budget approvals (even reducing costs from MYR 10M to MYR 1.6M while expanding our impact), and led cross-stakeholder alignment between PROTON, GEELY, and external distributors. The budget is not expensive, but I want to emphasize how I basically operate in a lean resources environment but still can deliver the results. If everything goes well, PROTON export vehicles will have a major upgrade to the connectivity elements—the first time this is done in a long time—a major transformation achievement considering a lot of internal and external challenges.

- International Product Strategy and Development: I drove product-market fit studies, lifecycle marketing, and coordinated global launches across all regions. We are now also expanding to new regions, possibly twofold of our current markets.
- NEV Export Leadership: I led PROTON's first-ever global NEV export strategy for BEV, HEV, plugin hybrid EV (PHEV), and range extended EV (REEV) models, working directly with the GEELY Team to ensure regulatory compliance and competitive positioning.

This multi-phase transformation that I mentioned, if executed well, aims for a USD three billion revenue impact within the next decade, considering the market potential.

Beyond just strategy and numbers, I take a lot of pride in building people, specifically automotive experts in Southeast Asia. Despite having a relatively junior team, I deliberately invested time to train them into elite global automotive business strategists—blending my Nordic (strategic, disciplined) foundation with ASEAN (emerging, high-growth) frontier market execution playbooks. I also dived and got myself into the regional ASEAN automotive ecosystem by connecting with automotive people from Malaysia, Indonesia, Singapore, and even Vietnam.

For me, real leadership is not about being insecure or hoarding credit—it's about growing the best team around you. A great team doesn't threaten a real leader; it pushes them higher. And when you build a strong team, they don't just work for you—they believe in the vision with you.

That's the spirit I bring to my journey today—and into the future.

And as of now, while this book is being revised, I have managed to lead the foundational task force at the working level, helping my leader to set up PROTON International Corporation—the subsidiary of PROTON created to operationalize our international strategy, which I co-devised with my Head of International Division. The company was just incorporated in June 2025, marking another task I never expected to be exposed to, but I took the challenge to learn and grow.

This is essentially about building a startup within a legacy company, inside a global conglomerate, to transform the business model from a domestic-focused setup toward one with a stronger international business emphasis.

Key Moments, Lessons, and Pivots That Shaped My Professional and Leadership Journey

Throughout my career, various pivotal experiences have played a crucial role in shaping my leadership approach. At Sensible 4 in Finland, I was exposed to the niche and deep-tech world of autonomous vehicle technology. This experience gave me firsthand knowledge of how to bring new tech to global citizens and the complexity involved. It is not just hype, as some tech pundits claim. The challenges we faced in scaling our AV stack emphasized the value of teamwork, stepwise scaling, patience, and respect for each other. It's easy to get caught up in stress, but it's important to remember to look at our colleagues as human beings. These lessons laid the foundation for my leadership style, one that is human-centric in the commercialization of emerging technologies.

In Sweden, while working with ZEEKR, I found myself in a fast-paced, dynamic environment that demanded agility and entrepreneurial thinking—but this time, within a corporate setting. Although ZEEKR presented itself externally as a startup and tech company, internally there were the political dynamics typical of any large organization. This exposure proved invaluable for my growth.

Reflecting on my first ten years, I now realize that the intense challenges were not as damaging as they might have seemed—or at least, not as I felt at the time. Professionally, this role sharpened my ability to execute quickly while maintaining a clear strategic vision. As we navigated the complexities of next-generation mobility solutions and global digital product integration, I learned to balance speed with strategic clarity—an essential skill in today's rapidly evolving markets.

Returning to Southeast Asia, leading PROTON's global strategy from Malaysia has been an incredibly valuable experience in scaling a domestic brand to international markets. Managing vehicle exports, particularly EVs, across diverse regions has reinforced my belief in the power of combining local insights with global aspirations. My current role is not just about "product management." It is actually about leading group transformation from the perspective of global business strategy and international products. And if I do not document this properly, people will inevitably undervalue it.

That's why I believe that when we work in a rapidly changing environment, we must take the time to understand the actual value we are delivering—and document it clearly. Only then can we truly assess how much we are worth. For me, this book is how I plan to document that journey while inspiring others along the way. Regardless, I am grateful for the experiences these three companies have given me, providing a comprehensive understanding of what it means to commercialize new technology and a global international business perspective, from the Global North to the Global South. The journey has not been easy, and it continues to challenge me. However, this rapid growth has provided maturity in my work. And I always try to remember that we are human; we go to the office to find food for our families, so there's no need to be arrogant. Don't think of yourself as too important when dealing with people. But also, sometimes, you need to be strategically authoritarian in certain cases to get your job done.

These diverse experiences have helped me develop a leadership style that combines different roles across continents and cultures, preparing me to drive transformative change. Although I don't have the "sexy Big Tech" tag on my CV, I believe these kinds of experiences early in one's career are more valuable, both for me and for the organizations I work with.

The journey has not been without its challenges—underestimation, undervalued, and unrealistic expectations. These, together with iterative feedback and stepwise progress over perfection, have taught me to build carefully within unstructured constraints, an approach I now consider essential in my leadership style.

Together, these lessons have honed my ability to lead with both resilience and foresight, adapting to challenges while keeping an unwavering focus on achieving transformative outcomes. I believe this is enough for this subsection, good enough for you to read and comprehend. And if you want to know more, you can always ask ChatGPT. As my previous boss in Sweden once said, "If you can say ten things, say two things." It's not about what you know; it's about saying less but making people understand more. Concise, precise, and sharp communication creates a lasting impression.

The Strategic Advantage of Multipotentialites in Global Mobility

Through my journey, I've learned that the automotive industry, and indeed any global industry, needs multipotentialites—individuals who can blend deep-tech expertise with strategic thinking and who understand both the big picture and the granular details. Multipotentialites bring immense value by bridging innovation and implementation, making them invaluable in complex global markets.

Throughout my career, each region and company I've worked with has taught me one dimension of leadership. But I've always brought a cross-dimensional perspective when working with them. Together, these experiences have shaped my view of global mobility and highlighted why diversifying skills is crucial.

The intersection of these experiences—technical, strategic, and cultural—has shaped my understanding of global mobility. As the industry moves toward more connected, electrified, and AVs, I've realized that the future of mobility isn't just about technology—it's about scaling that technology across different markets, cultures, and systems to create truly global solutions.

This chapter provides aspiring professionals with a strategic view of how building diverse skills across regions and roles can propel their careers. For business managers and recruiters, it will demonstrate the value of hiring individuals with broad competencies who are capable of tackling complex global challenges and leading across industries. Many of the insights shared in this book are only truly understood by those living through rapid change. Unfortunately, only a small portion of today's job landscape fully grasps this, though the shift will soon impact the broader global workforce. So, be ready—both employees and employers. Equip yourself.

Chapter 06

Transdisciplinary Strategic Leadership: Integrating Technical Expertise, Global Perspectives, and Lifelong Learning

So far, I've shared the background of my journey, the changes in the industry, and a detailed look at the lessons I've learned from more than a decade of navigating this fast-evolving landscape. I've taken you through key moments that shaped my leadership style, from honing technical skills to embracing international experience, and from working in startups to corporate environments. But now, I want to shift the focus.

I want to dive deeper into what it truly takes to stay relevant in this ever-changing career landscape. What does it mean to be future-proof, to lead with intention, and to remain adaptable in a world that's constantly shifting? I've come to realize that the road to continued success requires reflection—deep reflection—not only on what you've done, but on what's needed to stay ahead.

This chapter, while a bit repetitive, is intentional. The ideas I share here are meant to prompt each reader to reflect on their own career journey and their own storyline. I don't want to create a blueprint for everyone to follow, nor suggest that there's a singular path to leadership. What I want is for you to use these reflections as a mirror to understand what is needed for your own success.

Of course, like the rest of this book, I won't make it too textbook-like. Instead, I will take my own example and share it as a lesson for you.

Technical Prowess and Lifelong Learning for Competitive Advantage

Throughout my career, technical expertise has been a cornerstone, and it continues to be central to my approach as I lead global automotive business expansion strategies today. Whether at Sensible 4, where I was involved with self-driving AV technology, at ZEEKR, where I worked on data management for a digital software-defined ecosystem for EVs, or at PROTON, where my understanding of deploying EVs and ICE with connectivity in frontier and emerging markets shaped top management decisions, technical knowledge has always been at the core of my work. However, being technically proficient is only one part of the equation.

In today's world, staying ahead requires more than just expertise; it demands continuous learning and adaptability. My key belief is to create solutions that are compelling enough to naturally attract market demand when the company pushes them—it's not just about a one-way push or pull but a mutual push–pull dynamic. This approach goes beyond merely transitioning from a technical engineering background to business. I have deliberately made an effort to deepen my understanding from a technical perspective by taking a variety of business and management free online courses during my time in Finland amid the COVID-19 pandemic and by pursuing an Executive MBA at the Gothenburg School of Business, Economics, and Law.

In my journey, certifications, conferences, side projects, and even moments of quiet reflection have played a crucial role in ensuring that my technical knowledge evolves in alignment with industry advancements. Whether learning about new digital technologies, potential substitutes in CASE mobility, complementary businesses for the next-gen automotive landscape, understanding geopolitical shifts, or refining pricing strategies, my commitment to lifelong learning has been a critical factor in staying relevant in an ever-changing landscape. This commitment to learning is what separates those who keep up from those who fall behind.

At GEELY, I've always taken a proactive approach, engaging with up-and-coming startups and speaking with their founders to explore potential solutions and identify synergies that could benefit our current plans. Being active on LinkedIn has been immensely helpful, granting me access to the right people and facilitating valuable conversations. Doing this consistently for years has fundamentally shifted the way I work and think as a professional.

Yet, this commitment goes beyond just technical relevance. It is about having the curiosity, humility, and adaptability to embrace new concepts, ask tough questions, and continuously challenge my own assumptions. These mindsets, I've found, have been essential not only for navigating the technical world but also for thriving in the complex global markets in which I operate.

I believe that someone without technical knowledge will struggle to lead an organization effectively in the era of disruptive transformation. Similarly, someone who only understands technology, without a broader business perspective, will face similar challenges. Leadership today requires the ability to bring together diverse points of interest, synthesize them, and create a unified vision for the future. It's about developing a comprehensive roadmap that integrates technical insights, business acumen, and strategic foresight, ultimately guiding the organization toward success in an increasingly complex and dynamic world.

Shifting Mindsets: From Projects to Products and Embracing International Experience

Early in my career, I worked on projects. Projects are great—they have clear start and end points, defined deliverables, and measurable outcomes. However, projects are often short term. When I transitioned into a role focused on products, that's when everything clicked. Products are about long-term value creation; they're about building something that stands the test of time. And they require a different kind of leadership—one that understands the bigger picture and prepares the right infrastructure and processes to ensure continuous, sustainable deployment. This also involves considering the back-end stream, including multi-tier supply chains, to maintain alignment with broader business goals.

At PROTON, when I led the first NEV export roadmap, I quickly realized that the true challenge wasn't just in delivering vehicles to new markets—it was about creating an ecosystem. It required thinking not only about the product itself but also about the entire infrastructure that supports it. The roadmaps I developed were not purely technical; they were strategic, encompassing market insights, cultural adaptation, regulatory compliance, and future-proofing the business. This product thinking is essential for anyone who wants to remain relevant in today's market.

We engaged with potential partners, and due to lean resources, I held numerous internal discussions with partners across the globe through Teams. I synthesized this output to create our own emerging and frontier economy NEV export playbook. While this report is extremely valuable externally, it required significant internal buy-in for the change to take hold. As I said, my focus is on creating something that will not only bring success to the company but also foster my own long-term growth. This mindset is something that you, too, should adopt.

International experience has been crucial in shifting my perspective. It has taught me how to navigate ambiguity, adapt to diverse cultural and business contexts, and learn from multiple perspectives.

Each country I've worked in—whether in Malaysia, Sweden, Finland or elsewhere—has added layers to my strategic thinking. These layers have allowed me to better understand the global business ecosystem and how to drive success in a world that demands speed, agility, and cross-cultural collaboration.

One of the competitive advantages I now possess is the ability to understand that business is, at its core, about making money. If I were still in Europe, I might not have gained this clarity. Let me give an example. The top leadership wanted to push for exports to Nepal, an emerging market. Instead of crafting a full-fledged strategy tailored solely to the Nepalese market, I introduced a trimmed-down strategy designed for emerging markets. This internal "blue ocean" strategy leverages our position as a carmaker in Southeast Asia to target emerging markets—an approach that is rarely pursued by others. This perspective has provided me with greater openness and clarity when devising business strategies. I doubt that, had I remained in Europe, I would have understood how to deliver results with such scattered, lean resources and limited data.

Additionally, this experience has shaped my belief that, if you seek to develop specific skills, you must look for the right organization—not just focus on the badge it carries. For me, I wanted to learn about startups and scale-ups, and how they operate sensibly. If I had gone to a larger spin-off by a corporate giant, I would not have had the opportunity to truly live and understand the required skills, since those environments are often more diluted, with corporate players calling the shots. The atmosphere at Sensible 4, for example, provided me with a Silicon Valley "HBO sitcom" vibe, which has been invaluable. Similarly, if I wanted to learn about global expansion, PROTON is the right place. It's not a giant already operating globally; it's a growing company with an exciting future ahead. On the other hand, if I wanted to learn about international business governance, I would go to a larger, more stable corporation already established in every market.

It's all about understanding what skill sets you can gain from each company.

Startups and Corporate Perspectives: Balancing Agility and Stability

Startups and corporations are often seen as two extremes, but in reality, there's much that both worlds can learn from each other. Reflecting on my experience, I spent time in a fast-paced, high-agility environment at Sensible 4, where decisions were made quickly, the pace was rapid, and the structure was minimal. It was exhilarating, yet chaotic. Transitioning from that into the corporate world at ZEEKR and PROTON in GEELY, I was confronted with the scale, process, and politics of larger organizations. Corporations face their own set of challenges—complex systems, slower decision-making, and the weight of legacy processes.

What I've come to understand is that the best leaders know how to navigate both environments. In the corporate world, it's easy to get bogged down by processes, but I've learned to bring the startup mindset into corporate spaces. By injecting innovation into larger systems, I've been able to drive change from within. But I've also learned when to conform, when to push back, and when to challenge the status quo. This balance—knowing when to be agile and when to maintain stability—has been crucial to my success.

The Hidden Power of Startups: Lessons from the Trenches

Looking back at my career, one of the most defining experiences came during my time at Sensible 4, a fast-growing startup in Finland. It wasn't just the rapid growth or the exciting challenges that shaped me—it was the foundational work, the quiet, often unseen effort that helped drive the company from its early stages to a ~€20 million collective total of funding and revenue stream in just a few years. In a startup, nothing is handed to you on a silver platter. It's all about rolling up your sleeves and getting things done.

People often ask how I advanced so quickly in my career, and the answer is simple—growth isn't always visible, but it's built on the foundation of hard work, adaptability, and the ability to navigate chaos. Being part of a startup at such a pivotal stage gave me invaluable lessons that have become the cornerstone of my career:

- Structuring from scratch: I was directly involved in scaling the organization from 20 employees to 100 in less than a year. It was an intense, sometimes overwhelming process, but one that taught me the true meaning of organizational growth.
- Managing internal crises: Every startup faces what I call "internal earthquakes"—moments of intense pressure and firefighting. If you've been through it, you know it's not just about solving problems, it's about doing so with grace, under pressure, and without losing sight of the bigger picture.
- Cross-functional communication: Building transparent and open lines of communication across different teams is critical. In a growing company, silos are the enemy, and learning how to bridge gaps across departments has been one of my most valuable skills.
- Flexibility with limited resources: Startups often work on lean budgets, but that doesn't mean we can't achieve big results. It's about making the most of what you have and doing more with less—skills that apply well beyond any single company.

These are just a few of the powerful lessons I took from my time in a startup environment. Many people don't understand that while salaries might be smaller in a startup, the growth you experience—both professionally and personally—is far greater. Even if a startup fails, the lessons learned are invaluable. Failure often stems from leadership in many cases, and for those of us on the team, that means we walk away with a clearer picture of what not to do as we continue our careers.

This mindset, this constant evolution, is what prepares you for leadership positions later in your career. Startups might feel chaotic, but they offer a level of growth and perspective that's hard to find in corporate environments. Even when you fail, you've learned something critical that will help you succeed next time.

To anyone starting out or considering a move into the startup world: don't underestimate the power of this experience. Even in the face of uncertainty, it will teach you lessons that will elevate your career for years to come. If you're willing to embrace it, the rewards will be greater than you can imagine. To the future, and to the lessons ahead!

Writing as Your Public Portfolio: Building Reputation and Thought Leadership

One of the most powerful tools I've used to stay relevant in today's world is writing. Writing isn't just about sharing knowledge; it's about building your public portfolio and shaping your reputation. Every LinkedIn post, every keynote, every TEDx talk has allowed me to showcase my thinking and influence the wider conversation.

It's through writing that I've built a reputation as a thought leader, sharing insights that not only help others but also open doors for new opportunities. I urge others to embrace this tool. It's not just about sharing what you know; it's about demonstrating the ability to turn complex ideas into digestible, impactful narratives that resonate with others.

And it's not just about content creation—it's about visibility. Writing helps you build a professional voice, a space where you can share your vision and ideas with the world, positioning yourself as someone who isn't just reacting to the industry, but shaping it.

Chase Quality and Impact, Not Mere KPIs

When I first started my journey in the automotive and mobility sectors, I didn't envision the profound role that writing would play in shaping my career. But here I am today, reflecting on my publications and the impact they've had—impact that goes far beyond simple metrics like citations or reads. As of January 17, 2025, my work has garnered over 600 citations on Google Scholar, nearly 70,000 reads on ResearchGate, and more than 25 Scopus-indexed publications, with more than 30 co-authors from leading institutions across the globe. These numbers, while significant, are not the end goal. They are by-products of a larger purpose—pushing boundaries and contributing meaningfully to the ongoing transformation of the industry.

I remember, in the early stages of my career, when I was pursuing my PhD, I would look at researchers with hundreds of citations and think, "Wow, that person has made it." What I didn't realize then was that their success wasn't just about accumulating numbers; it was about the quality of their ideas and the boldness with which they presented them to the world. It was the substance that made them stand out, not just the metrics.

As I've evolved in my career, I've expanded beyond traditional technical writing. I've ventured into strategic management and cross-functional areas, particularly around the commercialization of emerging technologies. The act of writing has become more than just academic—it's a way to reflect on my ideas, solidify my thinking, and build my personal brand.

What I've learned along the way is simple: don't chase KPIs; chase quality. Focus on the ideas that matter, the problems that need solving, and the innovations that can change the game. If you focus on delivering value through your work—through thoughtful, meaningful contributions—the metrics will follow naturally.

Writing is your personal portfolio. It's your branding in the long run. No one else will take ownership of your story—you have to write it down. Do it consistently, boldly, and with purpose. The world needs your insights, your perspective, and your unique voice. And in the process, you'll create a lasting legacy in your industry.

Remember, success isn't measured by how many KPIs you hit—it's measured by the impact you leave behind. If you focus on delivering quality, the results will speak for themselves.

Take Control of Your Narrative

Writing also gives you the power to control your own narrative. It allows you to shape how the world perceives you, your journey, and your contribution to your industry. But it's important to remember that this must be done ethically. Authenticity is key. Write from the heart, stay true to your values, and be transparent. Don't fabricate stories or take shortcuts. The credibility you build through honest and insightful writing will be far more valuable than any image you could try to create artificially.

In the long run, writing is the tool that will help you define your personal brand. It gives you the chance to show the world who you truly are and what you stand for. But do it with integrity. People will respect you more for your honesty and insight than for any polished, perfect narrative you might try to present. The power of writing lies not only in the words you share but in the authenticity behind them.

Before I move forward, I want to acknowledge that some of the content in this book comes from ideas I've shared previously on LinkedIn. I've taken those insights, paraphrased them, and woven them together to align with the direction of this book. Some of these thoughts are simply too valuable to be left behind, and I believe they deserve a wider audience.

The Importance of Networking: Lessons from My Journey

One of the most important lessons I've learned throughout my career, especially since returning to Southeast Asia from my time in Europe, is the true value of networking. But let me clarify something up front: networking is not about simply collecting business cards or making superficial connections. It's about building sincere, meaningful relationships based on mutual respect, shared values, and, above all, merit.

Since coming back to Malaysia, I've had countless visits from people—colleagues, mentors, and even those I met through professional networks—who are genuinely interested in connecting not just because of what I do, but because of what we can learn and achieve together. This is a clear reflection of the strength of the network I've built over the years, one that wasn't created with selfish intent but through sincere collaborations and contributions. The power of true networking lies in relationships that go beyond the professional realm; they become friendships grounded in shared experiences.

What I've found is that when you consistently deliver value—whether in the tech space, automotive, or any other industry—and do so with integrity, others will take notice. Networking should never be one-sided transactional; it must be built on merit, trust, and the desire to help each other grow. If you invest in people, work with good faith, and deliver results, doors will open naturally. It's about fostering relationships where the exchange isn't just about taking, but about mutual benefit and long-term growth.

In my personal experience, I've seen this play out over and over again. During my years working in the European automotive sector, I established relationships that have proven invaluable in my current role. Many of the faces I connected with back then are still a part of my professional network today. Whether it's reconnecting with friends in AVs, having insightful discussions with colleagues across different continents, or meeting new people who share a passion for innovation—these connections have transcended the professional and become true friendships. The strength of your network is not in how many people you know, but in how deeply you connect with them.

Networking, at its best, is also about understanding the market pulse. Business development is far more than selling—it's about recognizing industry signals, spotting trends early, and anticipating the future before it becomes mainstream. When I networked across PROTON's global presence with our distributors and potential business partners, it wasn't just about finding partners. It was about deeply understanding what each market needed, identifying what would drive success, and adapting our approach accordingly. This is where networking goes beyond relationships: it becomes a tool for gathering valuable intelligence and positioning yourself ahead of the curve.

Networking for insights, partnerships, and influence is how you stay ahead—not just by attending events or exchanging cards, but by truly engaging, learning, and anticipating where opportunities lie before they become obvious. By staying informed and connected with the right people for the right reasons, you can make decisions that not only shape your own success but also contribute to the future of your industry.

The key takeaway here is simple: build your network on the foundation of value, sincerity, and trust. When you consistently show up, deliver quality work, and treat others with respect, the right people will naturally find their way to you. Relationships based on merit are the ones that last, and in this interconnected world, you'll soon realize just how small the global industry really is. The relationships you build will not only benefit your career but will also lead to new opportunities that you never imagined.

So, if you're looking to grow in your career—whether in the tech, automotive, or any other field—remember that networking isn't about the number of connections you make. It's about the quality of the relationships you nurture. Keep showing up, keep delivering value, and most importantly, keep making genuine contributions. The right people will seek you out, and your network will naturally expand in ways you couldn't have foreseen.

Own Your Narrative: Get Involved Beyond the Day Job

Over the past few years, I've kept a close eye on the wave of layoffs and mergers reshaping the tech-driven sectors of the automotive industry. These headlines only reaffirm what experience has taught me: transformation will not succeed if we rely on the same legacy mindsets that once brought success.

When the goal is to truly transform—when technology, customer-centricity, and software become the core pillars—you cannot simply "adapt" on the surface. You must strategically understand these domains at a deeper level and integrate them intentionally into your way of thinking, planning, and leading.

Leadership starts with insights. I've made it a point to not wait for a title to lead. Whether through strategy documents, NEV roadmaps, or export plans, I've shared my thinking and contributed to shaping the direction of the industry. Thought leadership isn't just about being visible—it's about being strategic, being authentic, and driving change.

This is why, since 2022, I've taken it upon myself to contribute to the industry's broader knowledge. I edited four strategic industrial reports in collaboration with global contributors, tackling critical themes like V2X, privacy, and product governance for next-generation mobility.

Today, these reports are available online, indexed in Google Books, and published through SAE International:

- **Privacy for Software-Defined Battery Electric Vehicles:** https://www.sae.org/publications/technical-papers/content/epr2024012/
- **Responder-to-Vehicle Technologies for Connected and Autonomous Vehicles:** https://www.sae.org/publications/technical-papers/content/EPR2023010/
- **Product Governance and Management for Software-Defined Battery Electric Vehicles:** https://www.sae.org/publications/technical-papers/content/epr2024025/
- **User Experience for Digitalized and Smart Cockpits and Cabins of Next-Gen Mobility:** https://www.sae.org/publications/technical-papers/content/epr2025006/

These projects were not assigned to me by an employer, nor were they linked to any formal promotion. They were passion projects—pursued outside working hours, with a long-term vision in mind.

This brings me to a key point: if you want to stay relevant, do not wait to be asked. Own your career narrative. Step into areas like standardization, regulatory frameworks, and industry-wide collaborations—where the real shaping of the future happens. Don't think of these activities as side projects; think of them as investments into your bigger picture. Opportunities will come, sometimes from unexpected places, and often not directly from your current organization.

True leadership isn't about simply executing what's handed to you. It's about seeing the next curve before others do—and building yourself into it.

Frontier and Emerging vs. Developed Market Strategy—Lessons You Only Learn by Doing It

Sometimes you just need to do it to understand. Over the past two years, I've gained a unique perspective on what it truly takes to devise business strategy for emerging and frontier markets compared to developed ones. This is a skill set that can only be understood by those who have lived it—those who have been on the ground, navigating the complexities of these regions.

In developed markets, like the EU, the infrastructure is already there. Even for a company new to the region, the frameworks, expertise, and support structures are established. The road might be bumpy, but there are clear signs, clear regulations, and partners ready to assist. For example, when Chinese EV players entered the EU, GDPR was a hurdle, but it was a known one, with solutions readily available from consultants and experts.

In emerging and frontier markets, the story is entirely different. Here, you're not just entering a market—you're co-building it. The ecosystem isn't fully formed, and data is often fragmented. As you create your go-to-market strategy, you are simultaneously laying the foundation that will make market participation possible for future players. The end result might look similar—launching a product, reaching customers—but the process is far more dynamic and uncertain. Strategy, foresight, and the ability to make sharp decisions are crucial because resources are limited, and ambiguity is a constant companion.

This approach of building as you go became central to my work. There was no pre-baked playbook, no clear-cut path—just constant innovation and adaptation to meet the needs of each market. Now, with exposure across the full spectrum of global markets, from the most developed economies to the least developed, I've come to realize the massive potential these non-obvious markets hold. But this potential can only be unlocked with sharp execution.

Execution isn't just about having a plan; it's about navigating the complexities of these markets with strategic readiness, something that's only visible to those who are in the thick of it. As I continue to share more about international business and market strategy, it's because I see these opportunities becoming increasingly critical in the face of shifting geopolitics. There's a lot to be gained if you approach these markets with the right mindset and agility.

As I reflect on this journey and share my lessons learned, I invite others who are building, investing, or exploring new regional–global opportunities to connect. Together, we can unlock the true potential of these markets with the right strategic approach.

When Fast Is Not Enough: The Power of Strategic Pauses During Global Growth and Transformation

In my journey through various organizations undergoing rapid transformation, I've learned a crucial leadership lesson: sometimes, slowing down is the most powerful way to move forward. The need for speed is almost always present, especially during periods of significant change. But the art of leadership during transformation often lies in knowing when to pause, reflect, and recalibrate.

During my time at Sensible 4 in Finland, we navigated several stages of growth. From a small garage-based startup to a post-Series A company aiming for scale, each phase demanded different leadership approaches. Yet, one constant was the pace—everything moved at breakneck speed. The urgency to scale, innovate, and keep up with the competition was relentless.

The same was true when I joined ZEEKR, a company that had just acquired CEVT. We were in the midst of a dramatic transformation, moving from innovation to commercialization and from a Chinese company expanding into the EU market. The speed was again unrelenting. We had to adapt quickly, breaking down barriers, and aligning our offerings to the market. But amid this high-speed transformation, I often found myself reflecting: Are we moving too fast? Are we focusing on the right things? Are we losing sight of the bigger picture?

This is a question I've carried with me throughout my career. Sometimes, it's necessary to stop—pause for a moment, reflect, and ensure that we are still on the right path. It's easy to get caught up in the rush, to chase deadlines and targets, but when transformation is at play, it's vital to ask: What exactly are we trying to achieve? Are we adding real value with every decision?

Throughout my career, I've worked in environments where we balanced business-as-usual (BAU) operations with the transformation efforts. In those moments, leadership requires more than just speed. It demands a sense of balance. A leader must not only drive progress but also create space for the team to reflect and align with the company's broader vision. That balance is not always easy to maintain, especially when there are daily challenges to coordinate, operations to sustain, and new strategies to execute.

True leadership in times of transformation requires humility and clarity. Leaders must listen carefully to their teams, communicate openly, and, when necessary, slow down the pace to ensure alignment and prevent burnout. It's easy to blame lack of performance on talent, but more often than not, the issue lies in unclear communication or misaligned priorities. By fostering transparency and empowering teams to think critically, we can avoid missteps and stay on track.

Looking ahead, I'm particularly focused on scaling businesses across ASEAN and bridging ecosystems across borders—challenges that require both speed and thoughtful pauses. And while these efforts are ongoing, I continually return to the principle that, in the midst of all the urgency, it's the moments of reflection that make transformation sustainable. Sometimes, the best way to move forward is to stop, reflect, and recalibrate.

This lesson—validated by my experience and later reinforced by my academic learnings in my EMBA study—has become a cornerstone of my approach to leadership. In fast-paced transformations, we must remember that speed alone won't get us there. Sometimes, it's the strategic pauses that provide the clarity needed for lasting progress.

Building Global Impact: My Lessons on Growing Beyond the Conventional Path

At a recent session at the Ministry of International Trade and Industry (MITI) on the Malaysia-GCC FTA initiation and the Malaysia-Pakistan Economic Partnership Agreement (April 2025), I reflected on the importance of cross-border collaboration. From Malaysia's participation in a joint business and investment forum in the Southern Peninsula with Singapore (https://www.irda.com.my/js-sez/) to the strategic positioning of ASEAN within the global economy, one truth stands out: growth demands a broader perspective.

In my experience, particularly in international business, expanding beyond a narrow focus is essential to making an impact. I've spent years navigating the intricacies of both the developed Global North and frontier markets in ASEAN, where the opportunities are abundant but complex. I've seen how foreign players often struggle without the context of local intricacies, while some ASEAN companies, despite their talent, face challenges scaling globally. The truth is, to thrive, you must understand the big picture—which means taking on roles that push you outside your comfort zone.

This mindset has guided my approach, especially in the automotive sector. Working at PROTON, a company with a legacy rooted in Japanese manufacturer discipline but now a part of a Chinese conglomerate, has been a masterclass in blending traditional values with modern agility. Speed and adaptability are key in this new phase of growth, but they must be balanced with a deep understanding of market needs—something I learned by actively engaging with a diverse set of stakeholders, from OEMs to policymakers.

As I reflect on this journey, I'm reminded of the importance of *uchiawase*—a Japanese term meaning "gathering feedback and buy-ins." This philosophy has been central to my growth. To make an impact, you must actively engage with people from different walks of life, understand their needs, and build solutions together. Don't stay quiet in the corner; engage, learn, and build bridges across cultures and industries.

In today's shifting global landscape, this lesson is more relevant than ever. Whether you're looking to expand internationally or seeking new opportunities, understanding the broader context and embracing diverse perspectives will help you unlock your full potential—and ultimately, make a real, sustainable impact.

The Power of Networking and Side Projects for Career Growth

I've shared my thoughts on the importance of side projects before, but I believe it's worth revisiting, especially now, as many juniors and aspiring professionals have reached out to me for advice.

One key takeaway from my journey is that no one is responsible for your growth except for you. It's your vision, and it's up to you to make it happen.

The good news is that you can drive your growth through many avenues beyond your main role. One powerful way to do this is through side projects, especially non-competitive ones. These could include consulting in other industries, advisory roles for startups, networking, public speaking, thought leadership, or connecting with people whose careers you aspire to follow. It's these opportunities that not only broaden your horizons but also give you clarity on what to improve and where you're still lacking.

By engaging in side projects, you ensure that your skills don't become stagnant. You're making them transferable, and you're making sure they are recognized by those who can elevate you. More importantly, these projects keep you connected to a larger, dynamic ecosystem—one where you're not limited by your current environment.

One example I'd like to share is a great casual chat I had recently (Q1 2025) with Komal and Benedict from BloombergNEF about the vehicle electrification landscape in ASEAN. Ben reached out to me on LinkedIn, seeking to connect with key players in the EV and automotive communities in the region. I invited my team—Min, Sun, and Lim—to join the conversation, and we discussed everything from ASEAN's potential in shaping a global ecosystem post-tariff events, to the challenges of access to data in frontier markets, and perceptions of electrified mobility.

This discussion reaffirmed my belief that cross-border collaboration and communication in ASEAN are crucial for driving progress. These types of informal yet impactful conversations are exactly why I advocate for networking and side projects. They help you understand new perspectives, connect with people who can support your vision, and make you a part of an ecosystem that's shaping the future.

So, take a chance on something different, something non-competitive with your main role. It may just be the thing that accelerates your growth and prepares you for what's next.

Remember, growth happens when you push beyond the boundaries of your current role and take part in the larger conversations shaping the industries you care about. Whether it's through side gigs or casual discussions with key players, every connection and every project counts.

Let's keep the momentum going—if anyone is interested in fostering closer collaboration among ASEAN experts and business professionals in the next-gen economy, feel free to reach out.

Enabling Global Scale Starts with Internal Transformation

Sometimes, the biggest breakthroughs in career growth come from within. It's not always about waiting for the world outside to recognize your potential, but instead about leveraging the opportunities right inside your own ecosystem. This lesson has been a central theme in my journey, and I've realized that, much like any successful business, your career requires constant refinement, internal development, and strategic positioning.

Recently, I had the opportunity to attend TalentCorp's internal hackathon, where the goal was to digitalize Malaysia's national talent management system. One of the teams was proposing a global talent platform with monetization potential, while another was designing an AI virtual assistant to streamline integrated services. I couldn't help but see parallels between their innovation-driven mindset and my own career strategy. It's clear to me now that you don't need to wait for someone to recognize your vision—you have to *actively shape it*.

This brings me to one crucial insight: your career is your project. Much like a product undergoing continuous refinement, you must take ownership and approach it strategically. When I started out, I didn't expect anyone to understand or appreciate my path. I was aware that not everyone could see the transdisciplinary approach I was taking, especially in a world that still often thinks in silos. However, I trusted that the right people—those who are visionary and forward-thinking—would eventually see the value in what I was doing, and they did.

Take, for example, my involvement in TalentCorp's TF50 program. While many may not have noticed my journey from Malaysia to Europe and back, the government, in their forward-thinking approach, saw the potential. They supported my involvement in this initiative, pushing me forward as a member of this prestigious group. Having side gigs, networking, and engaging with the right people at the right time allowed my growth to expand beyond what was happening within my immediate environment. This goes to show that sometimes, your internal contributions might not be recognized immediately, but having a strong external network can act as the catalyst for scaling your impact globally.

This lesson was further reinforced by a conversation I had around Summer 2023 with a career coach from the University of Gothenburg, who told me, "You're incredibly proactive about your career." That stuck with me. Being proactive means treating your career as a *growth project*, not a passive experience. It's about understanding that success doesn't just happen. You have to actively curate it, like a product being refined for the market. You need to identify what people resonate with and market yourself accordingly. It's about storytelling—sharing your narrative in a way that positions you as the solution. But, and this is important, it's about changing the paradigm of how you brand yourself. It's not about exaggerating or fabricating details; it's about presenting yourself in a way that connects with others and demonstrates your value.

In this fast-paced world, no one will come knocking on your door, asking how they can help you succeed. You have to tell them why they should care, why your skills, your experiences, and your vision matter. When you can do that—market yourself as the solution to their challenges—that's when the doors begin to open.

So, remember: handle your career as you would a high-growth business project. Be proactive, treat it like an ongoing development process, and never stop marketing your value. The right people will eventually see it—and that's when the world begins to take notice.

Coaching the Next Generation

Today (Q1 2025), just before writing this part of the book, I had the privilege of speaking to a group of PROTON young talents, offering an overview of our international business. Instead of relying on slides, I chose to have a candid conversation with them. In the brief time we had, my goal was to understand their motivations and give them a glimpse into the broader landscape of the international automotive industry.

I emphasized that we are living in a time of profound change—not just in terms of technology, but also in managerial style and employee demographics. The reality is, many millennials are now stepping into middle management roles, while Gen Z is quickly following suit, already making strides in junior leadership positions. The critical question we must ask ourselves as leaders is this: Should we lead this new generation in the same way we governed those before them?

In my view, next-gen leadership should be grounded in five core principles: transparency, vulnerability, flexibility, respect, and accountability. These are not just buzzwords; they are essential in adapting to a time of transformation. The way we lead, the way we engage with our teams, and the way we interact with partners must evolve to stay relevant and effective. After all, it's not just about steering the ship—it's about setting the course for a future where everyone, regardless of age or background, feels empowered to contribute.

One thing that has become clear to me over the years is that coaching others is one of the best ways to foster growth—both for them and for yourself. When you invest time in developing the next generation, you're not just passing on knowledge—you're learning alongside them by listening to their questions. By teaching, guiding, and mentoring, you not only help others unlock their potential, but you also enhance your own. This dynamic of giving and receiving makes you a more effective leader, sharpens your skills, and broadens your perspective.

To retain and nurture the best talent, we must also embrace this mindset of continuous growth. The leadership we offer today should empower the future of the industry, not just for today's challenges, but for the next era. Coaching and mentoring are not just acts of generosity—they are strategic investments in the long-term success of our teams, organizations, and industries.

I've always been happy to connect with and coach the next generation of talent, especially my fellow Malaysians. It's my belief that, even in small ways, we can contribute to the long-term development of the ASEAN region. By guiding young professionals through their early careers, we ensure that we're all moving forward together.

As you reflect on your own leadership journey, consider the impact you can have by investing in those around you. Whether you're guiding someone just starting their career or someone stepping into their first leadership role, remember: coaching is not just about teaching others; it's about shaping a future that's more capable, more resilient, and more innovative.

My First Decade in Automotive and Mobility Software: Toward Software-Defined Vehicles—A Brief Remark

That is the title of my video when I delivered the speech.[1] Receiving the SAE Young Industry Leadership Award in 2023 was an unforgettable moment for me—a proud milestone in a journey that started in my home capital of Kuala Lumpur, Malaysia, and has taken me to places I never imagined. Standing in front of North American business leaders in Detroit, a Southeast Asian native working for a Chinese automaker in Europe, I couldn't help but reflect on the path I've traveled and the lessons I've learned along the way. My career, still in its first decade, has been an incredible ride that has shaped both my professional and personal growth.

When I look back, I see a young professional, once mesmerized by the twin towers of Kuala Lumpur, who later became part of a team that built the first AV prototype in Malaysia. From those early days in R&D in AD and extreme weather applications to leading strategic planning in future mobility, my journey has been full of challenges, collaborations, and learning. But it was never a solo effort. Achievements, like the Finnish Engineering Award we received in 2020 for our work in AD development, were the result of collective efforts—efforts that spanned continents, cultures, and industries.

One of the biggest lessons I've learned over the years, especially as we move toward SDVs, is that collaboration is paramount. The future of the automotive industry is not just about hardware—it's about the seamless integration of hardware and software. In this context, the value we create will be derived from how well we work together as an ecosystem. No one can do this alone. Whether you're working in the heart of Silicon Valley or in a small engineering firm in Southeast Asia, the need for continuous collaboration is undeniable. And this collaboration must extend beyond just the automotive industry; it should include public organizations, private companies, and cross-industry

---

1 https://youtu.be/x6-Oprfs5Y4

partnerships. This perspective has been shaped by my own experiences across various countries and industries, and it has allowed me to approach future mobility solutions with a truly global mindset.

But as I stand here today, I'm reminded that success doesn't come without its lessons learned. One of the key takeaways I'd like to share is the importance of mentorship and teamwork. Throughout my journey, I've been fortunate to learn from the brightest minds, and I've had the privilege of leading teams across different regions. The value of mentoring young professionals, particularly those from underrepresented backgrounds like mine in Southeast Asia, cannot be overstated. If I can do it, they can do it too. I believe that the automotive and mobility industries need fresh perspectives and diverse talent to innovate, and it's our job to ensure those voices are heard and nurtured.

Finally, as I look ahead to the next decade, I remain focused on sharpening my skills, learning from my mistakes, and fostering growth in others. This award is not a destination, but a reminder of the journey ahead—one filled with continuous learning, adaptability, and a commitment to making a meaningful impact. For those who may be at the beginning of their own careers or navigating uncertain paths, I urge you to stay committed, embrace challenges, and always keep learning. The road ahead may be complex, but it is ours to shape. Together, we can lead the way into the future of mobility.

Thank you once again, SAE, for this recognition, and here's to the future—together.

Lessons for Managers and Recruiters

In a recent article I wrote on LinkedIn when I wrote this part of the book (2025), I discussed how CASE mobility revolution presents a unique set of challenges and opportunities in talent acquisition and development. One of the key takeaways was the importance of focusing on team composition over size. A well-balanced team that integrates software expertise with traditional automotive knowledge is essential for the development of SDVs. It's not enough to simply grow a team by numbers; diversity in skills and experience is the critical element that drives innovation without sacrificing safety and reliability. Managers and recruiters must focus on attracting talent that brings a broad range of skills, fostering an environment where multiple disciplines can collaborate to create the next-generation mobility solutions.

Furthermore, I emphasized the need for integrative roles like product managers and product owners to bridge gaps between engineering, safety, and business objectives. These roles are pivotal in preventing silos and ensuring that development stays aligned with both technical and strategic goals. It's not just about hiring technical talent; it's about creating cross-functional leaders who can communicate effectively across departments and drive innovation with clear objectives. In the fast-moving ACES space, managers and recruiters must look beyond just filling positions. They must seek out individuals who are not only skilled but also capable of taking on integrative roles that shape the product vision from conception to market readiness.

Summary and Conclusion—Staying Relevant, Shaping Your Own Storyline

As I look back on my career, I see a journey of continuous adaptation, learning, and leadership. But what I want to leave you with is this: don't just follow a path—create your own. Each of us has a unique storyline, and the key to staying relevant is the ability to reflect, adapt, and continually push the boundaries of what you can achieve. The lessons I've shared here are not meant to be a roadmap to copy, but a set of principles to help you shape your own journey.

The future is uncertain, but it's also full of opportunity. If you're ready to embrace change, challenge yourself, and lead with insight, you'll find that the future is yours to define.

Part IV

Key Actions and Importance of Multipotentiality

Chapter 07

Embracing Multipotentiality and Building a Career That Grows With You

My career wasn't a straight line—that became my greatest asset. Early on, I thought the struggles, the pivots, the uncertainty were weaknesses. But writing this book made me realize: the ability to integrate diverse experiences is the strength that shaped my leadership journey. It is actually indeed one of the proofs of one's strategic mindset, to combine very complex input into one storyline. Of course, there will be a lot of people who don't see it and might even sound skeptic and try to drag you down. Trust me, I have been there. But also trust me that the right people will see you if you find them in the right place.

Of course, we cannot expect to have extravagant outputs if we just have one decade of career. But the learning curve that we had, that's what will show us our potential.

Whether earning an EMBA in the Nordics, completing a PhD, or navigating startup and corporate worlds across different countries, every step that I built provided me a hybrid identity—one that allowed me to work seamlessly across engineering, business, regulatory, and strategic domains.

If you're reading this and wondering how to future-proof your career: multipotentiality isn't just a nice-to-have anymore. It's a necessity.

Why Now Is the Best Time to Embrace Multipotentiality

Industries are converging, and the lines between roles are increasingly blurred. Today, no one hires "just" a specialist without expecting adaptability. The value of a single, narrow skill set is diminishing. Technology is evolving at a pace that outstrips job titles and traditional career paths, requiring us to evolve faster than static career tracks. Companies now need more than just experts in one domain—they need integration thinkers. They require individuals who can seamlessly blend technology, market insights, product knowledge, and regulatory understanding into actionable strategies.

This shift underscores the importance of multipotentiality. It's not about being a "jack of all trades"—it's about agility. It's about the ability to pivot across disciplines and make connections that others might miss. It's about understanding the value of multiple skills and integrating them in ways that drive results.

As industries continue to converge, we've seen a wave of redundancy and layoffs, especially in the post-COVID-19 landscape. At this early stage of transformation, companies no longer need just specialists—they need individuals who can integrate complex topics and lead the charge. In short, companies are looking for leaders who can navigate uncertainty, orchestrate diverse functions, and drive cross-disciplinary strategies. They need someone who can connect the dots across various fields and steer the ship through the storm. This is where we, as multi-potentialites, must rise.

Now more than ever, we need to expand and diversify our skill sets. I've embraced this multifaceted approach throughout my journey, and it has been critical to my success. For example, I contributed to €20M+ in collective funding and revenue at Sensible 4 through investments, public projects, customer projects, and awards. I played a significant role in shaping the strategic foundation of ZEEKR's entry into the EU from China, in addition to influencing its innovation roadmap. At PROTON, I navigated the complex internal ecosystem, laying the groundwork for the company's first NEV exports in 2025. I also drove a broad internal transformation—covering processes, strategies, communication, frameworks, vision, and planning—to help PROTON evolve from a domestic player to a global competitor for the first time in its history.

Each of these milestones was not about mastering a single discipline but about connecting the dots across different fields. The ability to move fluidly between technical, strategic, market, and regulatory domains allowed me to identify opportunities, solve problems, and earn trust where others might struggle. It's this ability to integrate and blend various perspectives that gives multipotentialites their edge.

Looking back, I also believe that for us to progress, collaboration across sectors is essential—private sector, public sector, and people. I even wrote about this in Automotive News Europe. As a Malaysian, I would love to see Malaysia gain international respect for its contributions. Toward the end of 2023, I had the honor of being the keynote speaker at Supercharger Summit 2023, one of the main innovation events in Malaysia in 2023 (and possibly in Southeast Asia). Speaking in my personal capacity, I discussed how to navigate uncertainty in business innovation, industrial transformation, and leadership required for the new talent demographic and job landscape.

Thanks to Safuan Zairi and the MRANTI team for the invitation to such a significant event. It gave me the opportunity to share my experience with relevant stakeholders. I spoke about working in deep-tech, emerging tech productization, and scaling up software-driven products. At the event, I emphasized the importance of collaboration across all sectors if Malaysia is to make significant progress, particularly given the rapid pace of industrial transformation. I look forward to the continued progress MRANTI will make, and I strongly believe that, together with other stakeholders, their vision can be realized.

The reason I share this story is that transformation is no longer confined to regions like Silicon Valley; it is spreading globally. With the recent 2025 US tariff event, this trend will likely become the norm. As industries converge, and the speed of change accelerates, we must ask: What does this mean for us as talent? We need to prepare for a world where regional shifts and global transformations are the rule, not the exception.

Whether it's a startup, a multinational corporation, or any other organization, the demand for integration thinkers—those who can create strategies encompassing the broader picture—is increasing. This is why now is the best time to embrace multipotentiality. It's no longer a nice-to-have trait; it's a key driver of success.

Building Scalable Career Systems: The Key to Long-Term Success

If you want a career that scales, you need more than just skills. You need systems.

In my career, I've come to realize that systems are what give you leverage. Understanding these systems allows you to navigate through them and persuade the relevant stakeholders around you. Systems enable growth, scalability, and leadership. Without knowledge of the systems, progress will be slow, and you'll struggle to move beyond your current position. I trust you'll get the idea.

Here are some mechanisms I use to gauge where I am right now. But a caveat: don't treat this like a holy book. Just as we approach our work with flexibility, we should approach our careers in the same way.

Feedback Loops from Mentors, Executives, and Peers

Constant feedback is essential for any career. It keeps you grounded and ensures you're progressing in the right direction. By actively seeking feedback from people at various levels—mentors, executives, and peers—I've been able to adjust my approach, refine my leadership, and make smarter decisions. To be honest, I even ask my direct reports for feedback on how I'm doing as a leader. It's essential to get a full picture from all angles.

Brand Building Through Public Speaking, Thought Leadership, and Visible Project Delivery

Developing a personal brand isn't about self-promotion—it's about demonstrating your credibility through action. I've consistently built my reputation through speaking engagements, thought leadership, and delivering visible, impactful projects. This creates a cycle of trust and influence that propels career growth.

Of course, a caveat again—some minority of the people will dislike this approach. Maybe jealousy? But great leaders and good people will like it and support it. For me, as long as you deliver at work, what you do outside of work is not others' problem. By engaging in these activities, I've managed to build my career not only internally but also externally.

I position my thought leadership not through loud marketing, but through execution clarity and credibility, focusing on results rather than empty words.

LinkedIn: Visibility Beyond Connections

LinkedIn is not about making friends, but rather about telling the world that you exist. It's about reaching recruiters, potential partners, and even customers. Keep posting on LinkedIn—share your insights, updates, and progress. But, and this is important, make sure you bring value. Don't post just for the sake of likes or trends. Don't share content that's simply copied from ChatGPT or other generic sources. Your network is looking for value, so deliver that.

Document your insights publicly through articles, talks, and social media posts. This helps build your credibility and positions you as a thought leader in your field. Don't wait until you feel "ready"—start now and refine your message along the way.

Focus on clarity and credibility, not on quantity. Quality thought leadership will always outweigh trying to be everywhere at once.

Preparing for the Long Game

Finally, when devising a plan for your career, accept that multipotential careers are messy and nonlinear. They don't follow a traditional path, and that's perfectly fine.

You need to build systems that help you manage ambiguity. From mentorship to a strong network, ensure you have the right support structure to continue growing.

Play long games. Focus on impact over titles and prioritize lasting influence. The journey may not always be straightforward, but by managing the systems that underpin your career, you'll be positioned for long-term success.

Challenges Ahead: Why Staying Relevant Is a Moving Target

The reality is that the mobility industries are fragmented and software-defined, and global markets are shifting faster than corporate structures can keep up. Ambiguity and change have become constants, not exceptions.

Your job is not to "arrive" at a perfect career. Rather, your job is to keep building systems, keep evolving your identity, and stay intellectually curious and strategically grounded.

This is what I'm focusing on today at PROTON:

- Leading global scaling initiatives.
- Shaping foundational systems that will outlast my direct role.
- Staying adaptable for whatever the next frontier requires.

And I invite you to build your career the same way—not as a static resume, but as a dynamic, evolving, high-impact portfolio of capabilities.

Summary

The future belongs to those who can connect across domains, build systems for scaling themselves, and navigate challenges with curiosity and resilience. Your journey will not be linear—that will be your greatest strength.

In this chapter, I've shared how my career has been anything but a straight path and how each pivot, struggle, and moment of uncertainty has shaped my leadership journey. The ability to integrate diverse experiences across engineering, business, regulatory, and strategic domains has been a key strength. It is not about mastering one field, but about connecting the dots and creating impactful, cross-disciplinary strategies.

For both employers and employees, this chapter highlights why multipotentiality—the ability to excel in multiple areas—is no longer just a "nice-to-have." As industries converge and roles become increasingly fluid, companies need individuals who can navigate ambiguity, adapt, and innovate across sectors. Specialists are no longer enough; integration thinkers are now the key drivers of success. This shift represents a significant transformation in the job landscape, where the demand is for leaders who can steer through complex, fast-changing environments and make connections others might miss.

For employers, this means looking for candidates who bring a diverse skill set, intellectual flexibility, and the ability to lead and collaborate across multiple functions. For employees, it is a call to embrace multipotentiality—combining various strengths and continuously evolving. As industries grow more complex, your ability to integrate experiences and lead across different domains will set you apart as a high-impact professional, capable of shaping the future of your company and career.

In today's landscape, transformation is not confined to specific regions or industries; it's global, and the need for dynamic, versatile leaders is more pressing than ever. Whether you're navigating a startup or a multinational corporation, success lies in building systems that support long-term growth, embracing adaptability, and positioning yourself as a leader who can manage the complexity of tomorrow's job landscape.

In conclusion, the key takeaway is simple: to succeed, focus on building systems for scaling your career, stay intellectually curious, and embrace the power of connecting diverse knowledge areas. The career of the future is not about linear progression—it's about continuous evolution, adaptability, and leadership across domains.

Chapter 08

The Future—Beyond the Horizon of Today

Looking ahead, the future of industries like automotive, tech, and beyond isn't about incremental improvements. It's about disruptive shifts driven by multipotentiality and cross-functional skills. In my career, I've always been a firm believer that to remain relevant in this ever-evolving landscape, you must develop the ability to think beyond your core expertise.

The automotive industry, where I've spent much of my career, is a sector in transition—from ICEs to EVs, from ownership models to shared mobility, from manual driving to AD. But the true disruption isn't just about technology or product. It's about ecosystems that transcend traditional boundaries. The rise of smart cities, data-driven vehicles, and integrated mobility solutions requires a blend of expertise in product development, market strategy, technology, and policy.

For a long time, I focused on this: understanding how these industries intersect. Looking back, I was never just a product strategist or just a market entry leader. The more I worked, the more I realized that next-generation businesses are those that can cross-pollinate ideas, embrace new business models, and—most importantly—bridge various industries together. The future isn't just EVs or AD—it's everything that connects them.

This realization is what's guiding my current thinking. What excites me isn't simply the future of one industry, but how these industries can collaborate and create new value chains. Imagine a world where automotive players collaborate with fintech startups to enable seamless vehicle financing solutions, where mobility platforms are integrated with healthcare systems for optimized patient transport, or where even a more decentralized, borderless trade in global markets is driven by smarter logistics systems. This is the power of convergence.

We are also seeing the change of demographics. If leaders in the industry don't want to change, I give it a few more years before they start struggling to lead their talents. Just look around—millennials and gen Z now occupy half of the workforce (if not more) in most companies. Compare their lifestyle with that of older leaders. Will the current way of managing, recruiting, and retaining talent be sustainable? That's why we, as talent ourselves, need to have diversified skills to weather this storm. Additionally, we are also seeing the potential for supply chains to increase in the Global South, another demographic shift driven by the latest trade war dynamics.

During my time in the EU, I realized that developing countries like Malaysia have an abundance of homegrown talent in tech, particularly in mechatronics and robotics. I've met brilliant Malaysian engineers with outstanding ideas. However, what might be lacking is the commercialization aspect—access to the broader tech ecosystem and the network needed to scale globally. My point is productizing emerging tech is not only about innovation but also about understanding market dynamics and commercialization. That's why, to really push innovation forward, we need more talent with cross-functional skills.

Furthermore, with the recent news on tariffs potentially impacting major tech players like the US, China, and other developed markets, I foresee tremendous growth potential in the cross-border tech ecosystem across Southeast Asia, North Africa, West Africa, the Middle East, and even Latin America—especially considering the young and highly educated tech talent in ASEAN capitals. I even believe that the actual potential market for emerging technologies' mass production will be in markets with a young population, such as those in ASEAN and the African continent.

One of the key learnings I've had over the years is that innovation cannot be forced. It needs to be nurtured organically, driven by collaboration within the community. The tech community and ecosystem need to be empowered. This means we need more organic cross-collaborations between private, public, and people-driven organizations. "Organic" is key here. This perspective of mine has been shaped by my nearly six years in the Nordic tech scene, particularly in Finland and Sweden. Similarly, we cannot force change upon ourselves in one day—take your time, and let it evolve slowly.

For years, I've been building my career with this understanding in mind. The path forward isn't just about following traditional playbooks. To truly lead, we need to think beyond silos, continuously develop cross-functional skills, and build networks that allow us to navigate these increasingly complex environments. And that's exactly what I'm looking to do next in my future. We are at a critical juncture, where industries must embrace cross-pollination and collaboration to thrive.

With major layoffs ongoing on the global stage, it's clear that diversification is key. As employees and talent, we must adapt, diversify, and become attractive to the evolving landscape. The need for agility, flexibility, and an expansive skill set has never been more critical. And, to be honest, especially with the 2025 tariff event, I foresee the rise of new markets that will have an impact on the entire talent and supply chain of the automotive industry. The way these emerging markets shape our strategies and supply chains will be pivotal in defining the future of the industry.

Looking back, the reason why I wrote this book is because I believe that the career landscape is currently quite tough out there. Every day, we hear about layoffs in the news. Since I'm the minority who has this kind of profile in the automotive industry, I feel a strong responsibility to contribute to the advancement of the workforce and transform the industry. This book idea has gained further amplification, especially after receiving the SAE Young Industry Leadership Award in 2023 in Michigan, US. The award highlights my diverse international contributions in areas such as active safety and ADAS; automotive software and AD; automotive and mobility deep-tech startups; and leadership in automotive and future mobility standardization. This recognition is not just a milestone, but a reminder of how far we can go when we cross traditional boundaries. It reflects my passion for driving innovation through convergence, and I hope it inspires the next generation of talent to believe that their diverse backgrounds and skill sets can lead to success in a rapidly evolving world.

I am especially grateful to my mentors across multiple countries, including Zamzuri, Raksincharoensak, Limbu, Wong, Santamala, Saarinen, Olsson and many others, whose guidance has been invaluable. A special shoutout goes to Zhang, who has always believed in my potential and pushed me to grow. As Jim Farley, CEO of Ford Motor Company, rightly said at the event where I received the award, for every success, there is someone who nurtured and encouraged us along the way.

In sharing this journey, I hope to motivate young professionals like me, particularly those from minority groups. If I can achieve this, so can you. The world is borderless now, and opportunities for growth are abundant. I look forward to bringing the lessons, perspectives, and insights I've gained from my international experiences and implementing them into Southeast Asia. This is just the beginning. I'm excited for the next decade of my career—to sharpen my skills, learn from my mistakes, and leave an even greater impact.

To the future!

Key Takeaways that I want to emphasize from this chapter:

- Multipotentiality and cross-functional skills are the future of industries. The more diverse your expertise, the more valuable you are.
- The automotive industry is a perfect example of how systems and industries are converging—those who don't adapt to this ecosystem approach risk falling behind.

- The future isn't about isolated industries—it's about the intersections between them. From automotive to fintech, mobility to healthcare, convergence is where innovation lies.
- The 2025 tariff event and the rise of new markets will have a significant impact on the automotive industry's talent and supply chain, reshaping strategies and global operations.
- Leadership in automotive innovation is key to unlocking future opportunities, and young professionals can play a vital role in shaping that future.
- Emerging markets have an untapped pool of homegrown tech talent, especially in mechatronics and robotics, but commercialization and global scaling require stronger ecosystems and networks.
- The emerging markets tech ecosystem is poised for growth, particularly in light of shifting global trade dynamics, and fostering organic cross-collaboration will be essential to harness this potential.
- Multipotentiality enables individuals to thrive in an increasingly complex, cross-functional environment, where the ability to adapt, connect, and innovate across various industries will become the defining characteristic of tomorrow's leaders.

Chapter | 09

This Chapter Doesn't Matter, but the Publisher Said I Need One

Transformation sometimes requires silence and reflection.

In honoring certain traditions and perspectives in book publishing, this chapter has been intentionally reserved.

Nothing important happens here. Just saving you from wondering where Chapter 9 went. Please proceed to Chapter 10.

This is a placeholder chapter.

Chapter 10

It Is a Fast-Changing World—Only by Collaboration and Diversification of Skills Can We Weather the Storm

You might have noticed I skipped from Chapter 8 to Chapter 10 (yes, I know there is indeed a Chapter 9, but Chapter 9 is just a placeholder chapter). It was purely intentional—to show how fast the world is changing. And finally, after a couple of years, the book that was intended to showcase my one decade of career has managed to be completed. And we have reached the end of the book.

Let's recap again why I wanted to do this book.

Working in the disruptive industry, I don't know about other people, but the opinion on what it means to have a job is always changing. Seriously, I'm not sure about others, but I don't think this is a normal thing in other industries. In stable industries—for example, manufacturing, legacy sectors, or industries where the products don't go through much transformation—the meaning of work might be more constant.

Life and Mindset Changes

Initially, when I submitted the proposal to Sherry, I was in a different kind of life.

I was in a different kind of mind state and perception of life.

I was in a different kind of career environment.

I was in a different kind of continent.

And I foresaw that this book would benefit only a certain number of people.

But fast forward—I feel I am now in a different kind of private situation. Hopefully a better one.

Not only am I working in a different business unit, but I have also moved back to my country.

So the perspective is also changing.

Evolution of the Book

Initially, this book was supposed to have 18 chapters. But then, I think maturity has gotten the best of me. I started to prefer concise material to be delivered, instead of deploying fluffy or half-baked books.

I think that's also the reason why bands like Arctic Monkeys or Fightstar, for example, now deliver very brief albums—only ten songs per album. And I guess that's the standard for the music industry nowadays: only ten songs per new LP.

I guess that also influenced my decision because I want this book to be really attractive for the younger generation. And this younger generation has a different mechanism to get information.

For example, ChatGPT is the norm for them, and TikTok is the norm for them to get information. They are more visual than text-based, whereas Google was the norm for millennials and older generations. And maybe for the generation before that—baby boomers and some gen X—they are still, to some extent, typing www.google.com into the browser to go to Google.

But you get the idea. Time is changing, and the way we convey our messages also needs to change. Since the aim of this book is mainly the younger demographic (i.e., young talents), conciseness is important.

So, if initially it was 18 chapters, now the actual book has only ten chapters. Yeah, I know—you might try to count the chapters in the table of contents again—it is indeed ten chapters. I have managed to condense and remove what is not needed for this book.

I believe there are several things I want to reiterate before we finish this book.

Let's recap this.

Book Recap

In the first part of the book, I have presented my story and detailed career background.

In this part, you will see how unconventional my career trajectory and speed have been.

In the second part of the book, I talked about the evolving industry and career landscape—covering what is changing in the automotive industry nowadays and what the challenges are in the productization of automotive mobility software.

Then, I talked about the evolving career landscape and what this means for talents like us.

In the subsequent part, the lessons learned, I provided more detailed studies and examples—and some mementos of events or time snapshots of my career—but in the form of lessons learned.

After that, I shared what we need in order to survive the future career landscape.

In the following part, we discussed key actions and embracing multipotentiality—why we need to do it now—and how to utilize the knowledge from this book for your own perspective.

Initially, I wanted to include forewords by other authors or industry leaders. But I felt it was more important for this book to resonate deeply with its intended audience. I wanted it to be written in my own voice—with a tone of coaching and guiding, or if you are older, simply sharing—to motivate and connect. That's why I asked a few aspiring young professionals from diverse backgrounds to contribute the foreword. This way, the book reflects not only my voice, but also the voices of the next generation. I want people to understand that the world is changing, and you cannot just be complacent with what you have—whether you are a company, a talent, or a business leader.

You need to understand that the world is changing.

To survive, there are pillars you need to address.

I think it is quite straightforward. However, from time to time, I might deviate from the original trajectory—sorry.

Final Reflections: The Importance of Future and Mental Health

Anyway, my point is: again—we are now, without realizing it, already in the future.

COVID-19 has left us.

We are now seeing the new advances of emerging brands of carmakers in front of our eyes.

We also see the struggles of legacy carmakers.

The transformation is rapidly changing—regardless of what is posted on LinkedIn corporate pages.

Please make sure that, despite the transformation and business needs, we also take care of ourselves.

This is what I want to emphasize in this book—mental health.

One of the main transformations that happened since COVID-19 is that, with the rise of the new generation, we see that mental health issues have gotten more traction and attention.

For example, a lot of resignations happened during the COVID-19 time—some of them were partially because of mental health issues.

This is also what I emphasized in my new year post earlier this year—that I will not compromise my mental health. At the end of the day, only we can truly take care of ourselves. If we ever find ourselves too overwhelmed, it's important to stop, reflect, and then run again.

Below is the post that I posted earlier on LinkedIn this year with some paraphrasing. Actually, it is also written as part of this book, to reflect that my career is now entering 11 years.

Building on 2024—Readiness for 2025 and Beyond

(Note that this part is written on my LinkedIn as 2024 reflection, moving into year 2025)

As I reflect on 2024, it has been a pivotal year of growth, transition, and reaffirmation of the path ahead. Each experience, challenge, and milestone this year has reinforced one truth: to stay relevant and impactful, we must continuously evolve, learn, and protect our well-being.

2024 Highlights
1. Leading PROTON International's EV Export Foundation

This year, I had the privilege of leading the foundational work for PROTON International's e.MAS 7 EV export initiative to four emerging markets. Having previously contributed to ZEEKR's European market entry, I now possess a complete view of international EV business dynamics—from China-to-EU import strategies to export strategies from Southeast Asia to other emerging economies.

2. EMBA Graduation amid Global Relocation

Despite relocating across continents for 75% of the program, I successfully completed my EMBA thesis and graduated. This journey sharpened my leadership and strategic thinking and unlocked new levels of self-awareness about my potential, which I intend to further cultivate in 2025.

3. Staying Connected to the Tech World

Even while focusing on commercial and international market strategy for ICEs and NEVs, I maintained a strong foothold in technology:

- Delivered a keynote at VEID 2024 in Shanghai alongside NIO's EVP.
- Published two strategic reports with SAE International on Privacy and Product Governance for BEVs.
- Contributed as an invited online speaker to the IROS 2024 Workshop in Abu Dhabi.

2024 Knowledge Strengthened
1. Expanding across International Business Areas

This year allowed me to deepen my involvement in international strategy, marketing, product management, and business development—strengthening my versatility across multiple business disciplines.

2. Navigating NEV Export to Emerging Markets

Exporting NEVs to emerging markets requires more than traditional go-to-market strategies. It demands a nuanced understanding of readiness levels, regulatory frameworks, and local ecosystems—an area I was privileged to immerse in deeply.

3. **Developing Strategic Market Insights**

   Leading an in-depth study across the Middle East and North Africa (MENA), African RHD (right hand drive) markets, and the Indian Subcontinent sharpened my ability to identify growth levers in diverse economic environments—an insight that transcends the automotive sector.

2025: Non-Negotiables and Aspirations

Moving into 2025, there are things I will no longer compromise:

- **Mental Health:** I will prioritize environments that respect and protect mental well-being. No ambition is worth sacrificing sustainable, healthy growth.

From these 2024 experiences, I offer the industry—not limited to automotive—a unique combination of strategic insight into international business expansion, emerging market dynamics, technology integration, and leadership in complex, evolving environments.

Looking Ahead

2025 is a blank canvas. The foundation has been laid. I am ready for new challenges, new collaborations, and new opportunities to drive innovation across industries and borders.

The next chapter is calling—and I intend to answer it with energy, resilience, and a relentless commitment to growth.

Final Reflections: A Decade Completed, A New Journey Begins

The end of 2024 marked not only the close of a remarkable year but also the conclusion of the first decade of my professional journey. Reflecting on these ten years, I recognize how varied and rich the experiences have been—across industries, continents, and disciplines.

Yet, when all is said and done, I do not wish to be remembered solely for the titles or achievements. I hope to be remembered for the results I delivered and for the experience of working together—as a professional and as a fellow human being.

Looking ahead, I see the horizon widening. Given the breadth of my experiences and a strong belief in my own potential, I am preparing to engage in broader discussions—particularly around international business trade and enabling companies to expand into emerging and frontier markets.

In short, I aspire to co-shape the future of businesses that seek to go global. If this resonates with you, please keep in touch with me.

This book marks the close of one chapter and the beginning of a new journey.

Let's continue to sharpen our capabilities, uplift each other, and build a better future for our shared human society.

Until we meet again—thank you for reading.

Index

A

AD technology, 79
Adaptability, 56, 102, 103, 106, 117, 130, 136
Ageism, 8
Agile, 63–64, 81, 106
AI virtual assistant, 120
Alipay, 48
Android Auto, 47, 48
ASEAN automotive, 36
Asian-based technology, 48
Authenticity, 109
Automotive mobility, 82, 87
  career landscape, 74
  multipotentiality for future relevance, 81–82
  requirements for future-proof team, 80–81
  role of multipotentiality, 79–80
  understanding multipotentiality, 76–77
Automotive mobility software, challenges and productization, 60–61
  attracting, retaining and managing the right talent, 64–67
  complexity of transformation, 57–60
  ensuring transparency across organization, clear vision and strategy, 67–69
  need for standardized taxonomy, organizations and functions, 69–71
  software-defined products, mindset and way of leadership, 61–64
Automotive sector, 61
Autonomous driving (AD), 4
Autonomous mobility, 70
Autonomous vehicles (AVs), 5, 6, 44, 52
  algorithm, 20
  commercialization, 6
  planning and control software, 58
  productization, 27
Autono-washing, 55

B

B2B commercialization models, 91
Baby boomers, 50, 56, 147
Balance market demands, 58
Battery EVs (BEVs), 11
Biases, 8
Bluetooth, 47, 48
Board-level decision-making, 92
Business-as-usual (BAU) operations, 116
Business development, 111
Business ownership, 90
Business strategy, 11, 70, 113

C

Career, 15, 135–137, 139, 145, 147
  environment, 146
  evolution, 13
  experience, 13
  feedback loops from mentors, executives and peers, 132–133
  global expansion leadership (PROTON), 93–96
  growth, 79
  landscape, 45, 74, 101, 139, 147, 148
  LinkedIn, 133
  long-term success, 132–134
  pivot to strategic innovation (ZEEKR), 91–93
  plan, 134
  public speaking, thought leadership and visible project delivery, 133
  tech contributor to team lead, 88–91
Career journey, 4, 5, 17, 102
  acknowledgment section, 20
  business and corporate strategist, 18
  commercialization goals, 21
  engineering manager, tech lead and expert, 18–19
  experts, 21
  global experience, 33–38
  industrial research program, 20
  initial strategy for ZEEKR EU's digital roadmap, 29
  international professionals, 21
  Japanese R&D methodology and mindset, 20
  key achievements, 38–39
  LinkedIn profile, 17
  market expansion and product strategy leader, 18
  product management, 22–24
  public outcomes, 29
  recruiters and collaborators, 19
  Sensible 4, 22
  SOP, 30
  strategy, marketing and business leadership, 24–33
  systemic-level thinking mindset, 21
  work–life balance, 22
Change management initiatives, 92

© 2025 SAE International

153

ChatGPT, 44, 45, 48, 133, 147
China automotive, 37–38
Chinese brands, 53
Chinese companies, 48
Chinese EVs, 53
Chinese experts, 37
Chinese teams, 77
Climate change, 47
Cloud-based services economy, 48
Coaching, 121–122, 148
Commercialization, 44, 91, 115, 138, 141
  of digital service products, 7
  of emerging technologies, 32, 96, 109
  strategies, 29, 59
Community-building, 3
Connected, autonomous, shared, electric (CASE) mobility, 6, 7, 11, 14, 28, 44, 46, 49, 56, 91, 103, 124
Connected mobility, 79
Connected vehicles, 4, 5
Constant feedback, 132
Constructive feedback, 68
Control system, 75
Coronavirus (COVID-19), 12, 50, 51, 54, 102, 148, 149
Corporate politics, 54
Credibility, 109
Critical decision-making, 67
Cross-continental collaboration, 92
Cross-disciplinary collaboration, 91
Cross-disciplinary strategies, 130
Cross-functional communication, 107
Cross-functional skills, 137–139
Cross-functional work, 63
Customer conversion, 47
Cybersecurity, 47

D

Data privacy, 91
Deep-tech contributions, 91
Demographics, 46
Developed markets, 70, 114

Developing strategic market, 151
Digital banks, 48
Digital behavior, 51
Digital business, 51
Digital experience, 54
Digital-first electrified vehicle models, 44
Digital innovation, 79
Digitalization, 14, 77, 92
Digital marketing, 10
Digital payment, 51
Digital productization, 81
Digital transformation, 6
Direct-to-consumer (D2C) model, 49
Diversification, 139
Diversity
  expertise, 78
  lack of, 4
  workforce, 35
Diversity, equity, and inclusion (DEI), 8, 13, 14

E

Ecosystem, 26–28, 44, 46, 55, 57, 58, 76, 78, 88, 91–93, 95, 102, 104, 105, 137, 139
Electric vehicles (EVs), 4, 5, 44, 45, 53
  collect system data, 47
  growth of, 46
  mass production, 73
  transition, 46, 60
Electrification, 43, 44, 77, 79, 93
EMBA Graduation, 150
Emerging brands, 148
Emerging markets, 141
Emerging technologies, 4, 61, 81, 139
Entry-level internal combustion engine models, 4
EU automotive, 34–35
EURO 4 emission standard, 47
European data privacy regulations, 60
European market, 92, 93
EV export planning, 44
Experience, 152
  automotive networks, 6
  AV commercialization, 6

Chinese-based automotive conglomerate, 6
commercialization of digital service products, 7
digital transformation, 6
East's economic powerhouse, 8
global perspective and multicultural dexterity, 6
keynote speaker, 7
leadership, 6
market entry and product management, 7
merit, 8
next-gen mobility business, 5–6
next-gen mobility product and market strategies, 6–7
ownership, 8
physical appearance, 9
primary goal, 9
reality in China, 8
Sensible 4, 7
startups and scale-ups, 9
strategic value, 9
subject matter expert, 7
tangible results, 7
technical knowledge, 7
timing, 8
top conferences, 7
Export strategy, 69
External factors, 56
External societal and generational factors, 52–55

F

Finnish Engineering Award 2020, 38
Finnish technology, 88
Fintech, 48
First-gen leadership, 10

G

Gen X, 50, 56, 147
Gen Z, 121
Geopolitical events, 52
GIM Robotics, 88
Global automotive business strategists, 95
Global business corporate strategist, 92

Global business ecosystem, 105
Global business strategy, 18
Global credibility, 34
Global digital product integration, 96
Global expansion and corporate strategy, 94–95
Global expansion leadership, 93–96
Global expansion planning, 63
Global experience
  ASEAN automotive, 36
  China automotive, 37–38
  EU automotive, 34–35
  Japanese automotive, 33–34
  US automotive, 36–37
Global experts, 23, 26
Globalization, 77
Global markets, 138
Global mobility, strategic advantage of multipotentialites, 98–99
Global product export program, 93
Global product planning, 80
Global product strategy, 69
Global supply chain dynamic, 52
Google, 147
Google Books, 113
Go-to-market strategy, 114
Government support, 46
Green recovery, 52
Greenwashing, 55
Guiding, 148

H

Harmony Intelligent Mobility Alliance, 53
Hierarchical legacy leadership mindset, 61
Human behavior, 51
Hybrid skills, 83

I

Individual task, 67
Industrial experts, 47
Industrialization, 61
Industrial market, 91
Industrial transformation, 132
Industry disruptions, 13
Infotainment head units (IHUs), 4
Innovation work, 68
Intellectual flexibility, 135
Internal "blue ocean" strategy, 105
Internal combustion engine (ICE), 4
Internal crises, 107
Internal ecosystem, 78
Internal legacy debt, 58
Internal sustainability, 62
Internal transformation, 131
International business, 82, 117
  areas, 150
  strategy, 63, 69
International experience, 104–105, 140
International export strategy, 8
International market, 97
  entry models, 18
  expansion, 81
  strategy, 150
International product strategy and development, 95
Iterative feedback, 97

J

Japan Automobile Research Institute (JARI), 20
Japanese automotive, 33–34
Japanese culture, 20
Japanese R&D methodology, 20
Job market, 82

K

Knowledge-sharing, 3

L

Leadership, 27, 32, 53, 56, 81, 95, 103, 105, 112, 115, 116, 122, 125, 129, 135, 141
  dynamics, 53
  experience, 6
  legacy, 69
  NEV Export Leadership, 95
  open, 68
  software-defined products, 61–64
  strategic, 78
  style, 56, 89, 97, 101
Learning curve, 44, 75, 77, 129
Legacy leadership, 69
Life and mindset changes, 146
Lifelong learning, 81, 103
LinkedIn, 12, 17, 21, 25, 26, 36, 37, 65, 67, 88, 90, 103, 133, 148
Localization strategies, 69
Long-term business strategy, 89
Long-term market growth, 90
Long-term strategic planning, 92
Long-term value creation, 104

M

Macroeconomics, 81
Macro-level industry, 43
Market entry, 92, 149
  and business strategy, 94
  EU, 29
  fundamental, 29
  strategies, 18, 92
Market-entry barriers, 47, 53
Market participation, 114
Market strategy, 69
Mass public, 44
Mechatronics, 138
Mental health, 149, 151
Mentoring, 122
Mentorship, 134
Merit, 8
Mobile apps, 55
Mobility-as-a-service, 55
Mobility preferences, 55
Model predictive control, 75
Modern automotive mobility space, 77
Modern mobility, 83
MooVita, 62
Multi-country experiences, 78
Multidimensional capabilities, 74
Multifaceted approach, 131
Multi-phase transformation, 95
Multipotentiality, 76, 82, 87, 130–132, 135, 137, 140, 141, 148
  definition, 77
  for future relevance, 81–82

importance, 130
mindset, 80
multipotentialite talents, 80
rise of, 77
role of, 79–80
strategic advantage of multipotentialites, in global mobility, 98–99

N
Natural disasters, 47
Need-to-know basis, 63, 64
Nepalese market, 105
Networking, 110–113
NEV export leadership, 95
NEV export to emerging markets, 150
Next-generation mobility, 53, 91, 96
Next-gen leadership philosophies, 89
Nordic culture, 35

O
Official task, 89
Operational mindset, 67
Operation governance, 31
Organic cross-collaborations, 139
Original equipment manufacturer (OEM), 49

P
Personal crisis, 14
Polestar 2, 46
Post-tariff announcements, 31
Privacy, 47
Product alignment, 18
Product commercialization, 39
Product development, 25
Productization, 60
Product management, 22–24, 97
Product-market fit, 79
Product planning, 91
Product strategy, 29, 92
Professional and leadership journey, 96–98
Professional networks, 110
PROTON, 5–7, 18–20, 30, 63, 80, 93–96, 104, 105, 131, 134

PROTON International's EV Export Foundation, 149–150
PROTON's NEV Export program, 7
Public audience, 46
Public initiatives, 46
Public–private partnerships, 26

Q
QR codes, 48

R
Racism, 8
Real market needs, 72
Regulation, 77
Regulatory compliance, 59, 79, 91, 95, 104
Regulatory environments, 59
Researcher, 5
Reverse culture shock, 35
Revolut, 48
Robotics, 5, 138

S
SAE Young Industry Leadership Award 2023, 38–39
Scalability, 60
Second-gen organization, 10
Self-driving, 70
Sensible 4, 7, 22–26, 37, 39, 57–59, 62, 63, 76, 78, 79, 88, 90, 91, 96, 102, 105, 106, 115, 131
Sexism, 8
Shared driverless ecosystems, 44
Shared mobility, 4, 5
Skills, 78
Skill set, 72, 135
Small automotive market, 13
Smarter logistics systems, 138
Social media, 29, 50
Software commercialization strategies, 59
Software-defined products, 61–64
Software-defined strategies, 44
Software-defined vehicles (SDVs), 43, 44, 123–124
Software development, 11, 58

Software-originated leadership philosophy, 62
Software-originated tribal mindset, 62
Spotify, 48
Stable industries, 146
Standard operating procedure (SOP), 30, 37, 62, 63
Startups, 106
  cross-functional communication, 107
  flexibility with limited resources, 107
  managing internal crises, 107
  structuring from scratch, 107
Strategic go-to-market planning, 79
Strategic innovation, 91
Strategic leadership, 78
Strategic mindset, 81
Strategic planning, 13, 27, 28, 69, 91
Strategic role, 92
Subject-matter expert (SME), 79
Sustainability, 47, 51, 55
Sustainable business model, 58
Sustainable cash conversion cycle, 25

T
Tactic execution, 53
TalentCorp's TF50 program, 120
2025 Tariff event, 139, 141
Tariffs, 8, 31, 52, 53, 138
Tech community, 139
Tech ecosystem, 141
Technical expertise, 102
Technical knowledge, 103
Technical skills, 101
Technology, 130
Third-gen strategy, 10
Tier 1, 27
TikTok, 54, 147
Tokyo laboratory, 33
Toxic corporate politics, 54
Traditional career paths, 130
Traditional linear careers, 74

Transdisciplinary strategic leadership
  coaching, next generation, 121–122
  cross-functional communication, 107
  flexibility with limited resources, 107
  frontier and emerging vs. developed market strategy, 113–115
  global growth and transformation, 115–116
  global impact, 117–118
  global scale, internal transformation, 119–121
  importance of networking, 110–113
  managers and recruiters, 124–125
  managing internal crises, 107
  power of networking and side projects for career growth, 118–119
  from projects to products and international experience, 104–105
  software-defined vehicles, 123–124
  startups and corporate perspectives, 106–107
  structuring from scratch, 107
  technical prowess and lifelong learning for competitive advantage, 102–103
  writing as public portfolio, 108–110

Transformation, 4, 32, 51, 53, 54, 63, 67, 69–71, 132, 143, 148, 149
  complexity of, 57–60
  digital, 6

Trump, Donald, 53

T-shaped professionals, 77

U

US automotive, 36–37

V

Value, 9–10
  for employees, 11–12
  for employers, 10
  for general readers, 12–13

W

WeChat, 48
WHIM app, 55
Work dynamics, 34
Work from home (WFH), 50
Work–life balance, 22, 36, 88
Work philosophy, 50
World economy, 47
World Resources Institute, 46
Writing, 108
  chase quality and impact, 108–109
  narrative, control of, 109–110
  power of, 110

X

X as a Service (XaaS), 17, 43

Y

Younger generation, 50, 54, 146

Z

ZEEKR EU, 7, 29, 44, 59, 79, 93